GENDER AND POLITICAL ECONOMY
Explorations of South Asian Systems

edited by
ALICE W. CLARK

DELHI
OXFORD UNIVERSITY PRESS
OXFORD NEW YORK
1994

Oxford University Press, Walton Street, Oxford OX2 6DP
Oxford New York Toronto
Delhi Bombay Calcutta Madras Karachi
Kuala Lumpur Singapore Hong Kong Tokyo
Nairobi Dar es Salaam Cape Town
Melbourne Auckland Madrid

and associates in
Berlin Ibadan

© *Oxford University Press 1993*
First published 1993
First published in Oxford India Paperbacks 1994

ISBN 0 19 563461 6

Typeset by Rastrixi, New Delhi 110030
Printed at Ram Printograph (India), New Delhi 110020
and published by Neil O'Brien, Oxford University Press
YMCA Library Building, Jai Singh Road, New Delhi 110001

GENDER AND POLITICAL ECONOMY

Contents

Notes on Contributors vii

1. Introduction
   ALICE W. CLARK 1

I RELATIONS OF PRODUCTION/REPRODUCTION

2. Engendering Reproduction: The Political Economy of Reproductive Activities in a Rajasthan Village
   MIRIAM SHARMA AND URMILA VANJANI 24

3. A Woman Belongs to Her Husband: Female Autonomy, Women's Work and Childbearing in Bijnor
   R. JEFFERY AND P.M. JEFFERY 66

4. Analysing the Reproduction of Human Beings and Social Formations, with Indian Regional Examples over the Last Century
   ALICE W. CLARK 115

II CLASS, GENDER, AND THE INTENSIFICATION OF ECONOMIC FORCES

5. Social Classes and Gender in India: The Structure of Differences in the Condition of Women
   KALPANA BARDHAN 146

6. Patriarchy and the Process of Agricultural Intensification in South India
   PRITI RAMAMURTHY 179

7. Contradictions of Gender Inequality: Urban Class Formation in Contemporary Bangladesh
   SHELLEY FELDMAN 215

III GENDER IN-ITSELF AND FOR-ITSELF: ISSUES OF SOLIDARITY AND POLITICAL ACTION

8. Women in Development and Politics: The Changing Situation in Sri Lanka
AMITA SHASTRI   246

9. Organizations and Informal Sector Women Workers in Bombay
JANA EVERETT AND MIRA SAVARA   273

10. Development from Within: Forms of Resistance to Development Processes Among Rural Bangladeshi Women
FLORENCE E. McCARTHY   322

Index   354

Notes on Contributors

KALPANA BARDHAN is an economist with a Ph.D. from Cambridge University. She has taught at the University of California-Berkeley and has done consulting with the ILO in Geneva. Currently a translator of Bengali literature of resistance, she has published *Of Women, Outcastes, Peasants and Rebels: A Selection of Bengali Short Stories* (Berkeley, 1990).

ALICE W. CLARK is a historian and historical demographer with a Ph.D. from the University of Wisconsin. She has been a Research Associate in Demography and South Asia Studies at the University of California-Berkeley, and is currently a Lecturer in the Women's Studies Department there. Her current research is on women, children and reproduction in rural Gujarat.

JANA EVERETT is Professor of Political Science at the University of Colorado at Denver. The author of *Women and Social Change in India* (New Delhi, 1979), she has co-edited several other books on women and development. Her current research concerns urban Indian working women and reproductive issues.

SHELLEY FELDMAN is Assistant Professor of Rural Sociology and Associate Director, Program on Women and International Development, Cornell University. She is completing a book with Florence McCarthy entitled *The Gender and Development Matrix in Postwar Bangladesh*; her current research is on global economic restructuring.

PATRICIA JEFFERY is Reader in Sociology at the University of Edinburgh. The author of several books, she co-authored with Roger Jeffery and Andrew Lyon *Labour Pains and Labour Power: Women and Childbearing in India* (London, 1979). She is continuing research on childbearing women in Bijnor District.

ROGER JEFFERY is Reader in Sociology at the University of Edinburgh. He is the author of *The Politics of Health in India* (Berkeley, 1988) and has done joint work with Patricia Jeffery. His current research is on changing agrarian structures and relations of production in Bijnor District.

FLORENCE E. MCCARTHY is Associate Professor of Philosophy and Social Sciences at Teachers' College, Columbia University. The author of *The Status and Condition of Rural Women in Bangladesh* (Dhaka, 1978), she has also taught at Cornell University. Her current research is on environmental pollution and local knowledge in South Asian urban centres.

PRITI RAMAMURTHY is an Instructor in Anthropology and Women's Studies at Syracuse University. She has just completed a dissertation on irrigation systems in southern Andhra Pradesh, and has written several articles on the effects of changing agricultural technology on women.

MIRA SAVARA is a sociologist with a Ph.D. from Bombay University. She is the Project Director of Shakti, a feminist research collective in Bombay, and an independent consultant on women's issues. She is currently working on sexuality, reproduction, and AIDS in India.

MIRIAM SHARMA, an anthropologist, is Associate Professor—and recently Chair—of Asian Studies at the University of Hawaii. The author of *The Politics of Inequality: Competition and Control in an Indian Village* (Honolulu, 1978), she is currently working on a book on women's lives in a Rajasthan village.

AMITA SHASTRI is Assistant Professor of Political Science at San Francisco State University, and has her Ph.D. in Politics and International Studies from Jawaharlal Nehru University. She has written several articles on contemporary politics in Sri Lanka, the subject of her continuing research.

1

Introduction

ALICE W. CLARK

This volume of essays on gender as it operates in South Asia is situated at a midpoint between several areas of inquiry. Most broadly, it is intended to speak equally to, and form a bridge between, those engaged in gender studies on a world scale on the one hand, and those engaged in studies of South Asian societies and cultures on the other. Beyond this, it enters into the framework of a number of debates and attempts to refocus some of the questions within them.

We employ here a perspective that affirms the rootedness of gender in material conditions and arrangements, that is, a perspective in political economy. This is not necessarily to adopt a determinist view, nor to say that contributors affirm materiality as universally prior and causative; rather, it is forthrightly to claim the importance of studying material conditions and arrangements in the search for gender's historical and social context. In employing a perspective in political economy, therefore, we offer both a counter and a complement to the study of gender from the perspective of cultural criticism.

The authors, while all specializing in the study of South Asian societies, are from several disciplines in the social sciences, and they attempt to contribute to ongoing discussions within those disciplines, while at the same time embracing cross-disciplinary approaches. The studies in the volume both respond to and analyse specific empirical contexts personally known and deeply studied, and also attempt to theorize afresh based on new insights. For the most part, they occupy middle ground between efforts to theorize gender universally on the one hand, and the full-scale empiricism of many social science inquiries doing formal hypothesis testing on the other. In responding to specifically South Asian situations, they give rise to new questions that need to be dealt with through further and more closely honed hypotheses.

FEMINIST PERSPECTIVES AND GENDER STUDIES ON A WORLD SCALE

In the last decade, feminist scholarship on the international plane has both taken a quantum leap, and undergone a paradigm shift. Only a decade ago, caution about feminism as a Western intrusion, an anxiety about ethnocentrism, and a self-conscious sensitivity to cultural differences shrouded the most earnest efforts of 'Women in Development' specialists. Now, even with cultural sensitivity, it has become impossible to see the feminist movement in politics and scholarship as a purely Western intrusion. Women from many walks of life, in all corners of the globe— not a majority, but an historic few—are now consciously struggling with the inequitable burdens they and other women bear in relation to their livelihood, their political voice, their autonomy within their bodies and their very lives. These burdens are increasingly understood as being continually imposed on them by systems that support the subordination of women, both to men and in a larger sense.

The emergence of Third World women's movements has spurred new developments in feminist teaching and scholarship throughout the world. The last few years have seen a virtual explosion of books that wrestle with wrenching issues for women on an international scale. Two meant for wide distribution, via current courses in international women's studies, are representative of the new outspokenness. One is a dazzling demonstration of computer graphics called *Women in the World: An International*

Atlas (Seager and Olson, 1986), and another is a statistical handbook with a clenched fist entitled *Women: A World Report—A New Internationalist Book* (1985).

Correspondingly outspoken movements have occurred in the area of feminist scholarship per se. A vigorous literature has developed on gender from a political economic perspective in a world framework. This area of inquiry is of much more recent origin than South Asian studies. It has some of its roots in Marxian perspectives on development and underdevelopment, and in world systems theory. A bold attempt at an analysis of gender within a developmental world framework is attempted in Maria Mies's *Patriarchy and Accumulation on a World Scale* (1987).

A world perspective is gaining salience across many fields, whether indebted to similar frameworks and theories or not. Scholars operate out of a political economic perspective oriented towards world-wide changes whenever they attempt to understand social transformations flowing out of such world processes as the penetration of colonialism; the development of market economies; the historical coalescence of a world economic system; and the increased hegemony of capitalist relations everywhere.

The attempt to track the changing patterns of gender relations within these social transformations is carried out, in the first instance, at the level of the global economy. Supporting studies focus on the micro level in case-study format. Therefore, much of this literature is either general or comparative, or both. Most collections of essays on gender from this perspective contain both a few highly theoretical pieces at the global level, and case studies from a variety of locations around the world. Collections of essays on gender that look at realities falling within the realm of political economy sometimes contain individual contributions about South Asian women; an outstanding example is *Women and Property—Women as Property*, edited by Renee Hirschon (1984). What is needed is a more sustained focus on comparative realities within particular regions, in this case the South Asian region.

Universal or cross-cultural perspectives are sometimes

dominated by the drive towards, general theories of gender (Rubin, 1975; Kuhn and Wolpe, 1978; Young, Wolkowitz and McCullagh, 1981; Delphy, 1984; Smith *et al.*, 1984, 1988; Vogel, 1984; Redclift and Mingione, 1985; Acker, 1988; England and Farkas, 1986). Other essays and collections focus more particularly on Third World conditions, the classic progenitor being Ester Boserup's *Women's Role in Economic Development* (1970). This literature has mushroomed in recent years (Caplan and Bujra, 1978; Ward, 1984; Afshar, 1985; Young, 1986; Bisilliat and Fieloux, 1987; Mies, Bennholdt-Thomsen and Werlhof, 1988; Singh and Kelles-Viitanen, 1987). We attempt here to contribute to these global discussions from within a framework which focuses on a particular region with its own particularities, traditions and historical roots.

GENDER AND THE STUDY OF
MODERN SOUTH ASIAN SOCIETIES AND CULTURES

Although gender is a basic building block in any social formation or cultural tradition, its identification as an important subject for analysis in modern South Asian studies is fairly recent. My remarks on this topic are confined to studies of historically recent social conditions, and do not reflect on work done on the classics, or on Indian history before the eighteenth century.

Within the study of South Asian societies, a women's studies branch has developed, starting slowly in the seventies, building to a flood of literature by the late eighties. Explorations of women's condition in the region by South Asianists (Roy, 1972; Jacobson and Wadley, 1977; Dixon, 1978; Sharma, 1980; Omvedt, 1980; Everett, 1981; Miller, 1981; Papanek and Minault, 1982; Bennett, 1983; Caplan, 1985) have now been more than matched by an outpouring of writing published in South Asia by South Asian writers, editors and analysts, as the feminist movement has gained strength and voice in the subcontinent (Committee on the Status of Women in India, 1974; Jahan and Papanek, 1979; Das, 1986; Gulati, 1982; Kapur, 1982; Kishwar and Vanita, 1984; Desai and

Patel, 1985; Jayawardena, 1986; Desai and Krishnaraj, 1987; Jain and Banerjee, 1985; Chanana, 1988; Ghadially, 1988; Agarwal, 1988; Krishnamurthy, 1989; Sangari and Vaid, 1989). Two streams of literature, coming from feminist scholars dealing with women's condition from a distance, and those writing within the emerging context of an indigenous feminist movement, now interact and inform one another, and are no longer so distinct.

South Asian studies in general, however, have been somewhat resistant to finding gender a necessary category to include in broader inquiries. Studies of kinship, marriage, caste and village structure, for example, have traditionally been carried out without rethinking received assumptions regarding gender.

This has led to a lack of evaluation of the weight of social maintenance borne by women. Examinations of the cultures of the subcontinent have sometimes even obscured the agency of gender and the crucial role it plays within culture. For example, a view of kinship and caste as coded substance (Inden and Nicholas, 1977) makes it rather difficult to consider women as persons in their own right, or the gendered dimension of kinship. Similarly, scholars of marriage networks often discuss transactions in women without critical consideration of what this has meant for women's lives, or, more importantly, of what is necessary to maintain these relations. A feminist perspective on caste, or marriage alliance theory, of necessity will produce something new.

Indian social and political history, likewise, has been largely studied without inquiry into how life conditions have differentially changed for persons according to gender. For example, two otherwise excellent summaries of the socio-economic and political history of modern India (Charlesworth, 1982; Bayly, 1988) look penetratingly into the fabric of socio-economic life, but largely overlook women. This lack clearly should become impossible in future surveys.

Historical studies of class formation, and of class/caste articulation, have been advanced by the development of historical studies of people formerly hidden from the record, under the nomenclature of Subaltern Studies (Guha, 1982-7). These studies have, until recently (see O'Hanlon, 1988), been almost studiedly silent on the social category of gender, and the ways in which gender itself articulates with class and caste and adds another dimension to subalternity. An exception has been the trenchant cultural criticism of Gayatri Spivak (for example, in Basu and Sisson, 1986).

An area of South Asian studies in which awareness of gender has become inescapable in recent years has been the area of demographic behaviour. The remarkable masculinity of South Asian sex ratio data over the last two centuries, and its continuation today, has not escaped the attention of scholars from a number of backgrounds, not only demographers (Visaria, 1969; Mitra, 1978; Cassen, 1978; Chen *et al.*, 1981; Mason, 1984; Dyson and Moore, 1983), but also anthropologists (Miller, 1981, 1984; Das Gupta, 1987), historians (Clark, 1983, 1987, 1989; Greenough, 1982), and economists (Rosenzweig and Schultz, 1982; Bardhan, 1984; Cain, 1984; Sen, 1984, 1986; Agarwal, 1988). A report on a recent conference on the excess female mortality which is found in South Asia concluded that the issues were so densely interconnected that further examination would need to employ, preferably jointly, a variety of perspectives (Harriss, 1989).

An area of South Asian studies of much longer duration has been the social anthropology of family and kinship. Some specialists have always been attentive to gender relations (Karve, 1965; Goody and Tambiah, 1973; Mandelbaum, 1974; Jacobson and Wadley, 1977; Kolenda, 1978; Wadley, 1980; Bennett, 1983; Srinivas, 1986). Some have also been attentive to the political economy of these relations (Gough in Schneider and Gough, 1961; Mencher, 1983). An integrated perspective on kinship and gender, sensitive to issues of political economy, has been proposed in a recent anthropological collection (Collier and Yanagisako, 1987). (No South Asian studies are included, but the introduction and title essay are particularly relevant in pointing to the material dimensions of gender construction.)

Gender is considered in the study of the formation and functioning of labour markets, both formal and informal, as seen in the work of economists (Binswanger and Rosenzweig, 1984; Bardhan and Srinivasan, 1988; Jain and Banerjee, 1985), sociologists (Dixon, 1978; Omvedt, 1980; Lebra, Paulson and Everett, 1984), and groups of the above (Bjorkman, 1987; Agarwal, 1988). And the fact that women have been both acted for and active in political movements in the subcontinent is a topic that has been well documented (Wolpert, 1962; Heimsath, 1964; Everett, 1981; Liddle and Joshi, 1986; Jayawardene, 1986). A fuller than passing consideration of these topics, however, has often been segregated under the category of 'women's studies', and thereby marginalized by some claiming to work in the mainstream of modern South Asian studies.

The integration of gender into the mainstream remains tantalizingly out of reach. This marginalization of gender, which may perhaps be convenient in maintaining achieved male academic privilege, is only beginning to become unravelled through a number of new critical approaches.

GENDER AND CULTURAL CRITICISM

An influx of Foucaultian perspectives has swept through the humanities and social sciences in the last decade or so. With this development, coincidental with the rise of feminist perspectives and an awareness of gender issues, a new arena for gender studies has been opened up in the study of forms of consciousness. The deconstruction of social categories, when applied to the condition of women in South Asia, yields an acute awareness of the constructedness of gender in its various social and historical varieties.

Categories that have been socially constructed can be deconstructed and rethought, and the present collection adds to that effort in its own way. The perspective of cultural criticism is somewhat different than what we attempt here, however, and the distinctions are worth spelling out.

An excellent effort at cultural criticism of social constructions of gender in South Asia is found in Sangari and Vaid's recent collection (1989). Political economy clearly underpins and forms the context for the historical analyses of consciousness which are the subject of most of the chapters. That context is not itself, however, fully examined there, for the primary purpose of cultural critics is to deal frontally with consciousness itself, in the ways it has been shaped by the historical transformations of the context. This is extremely important; in a way, what we attempt here would not make sense to readers who had no awareness of the live states of consciousness that have emerged and changed and played their part in shaping the lives of women and men relative to one another, to society and state, and to the world order.

Yet with its strong emphasis on the role of socialization for class roles, via education, the media, the transformation of cultural messages, and the like, this approach perhaps overlooks the basic roots of class formation processes (see Sangari and Vaid, pp. 11-12). Class is not something just culturally produced. The level of development of the material possibilities underlies changing class relations or the rise of new classes, and is something that needs to be considered in relation to questions of how gender relations have adjusted. The element of materiality is the touchstone for the difference in approach.

The approach taken here certainly does not deny the socio-historical constructedness of gender, but rather deconstructs it in a different way. This difference need not be oppositional, as I will try to show.

The word *structure* has had a dual application in the social sciences (as evidenced in South Asian studies) for many decades. Both applications share a sense that what is called structural to a society is what is considered to be most basic to it. One sense in which structures are the most basic has led to examining the deeply-rooted cultural patterns that inform and direct people's living. This approach is concerned with text and subtext, the cultural code, the emic view. Another view is that material conditions, resources, possible means of producing livelihood, and their social distribution are basic. This approach is concerned with the material production and reproduction of society, and the social relations that further it, the etic view. South Asian studies have traditionally emphasized the former, more culturally oriented, approach.

We emphasize here a close attention to the second approach, acknowledging that it is both possible and necessary to deconstruct the duality itself. Gender studies are on the one hand returning to culture and upending our view of it, and on the other opening the socio-historical changes in production structures to new and critical questions. Feminist scholarship in many arenas is bringing about an awareness of the historical constructedness—and therefore impermanence—of the dualism between public and private spheres of life. This awareness offers an avenue for beginning to overcome some of the separation between the cultural and the political.

GENDER AND POLITICAL ECONOMY IN SOUTH ASIA

In bringing together world perspectives on gender and political economy with South Asian studies, this volume adopts an explicitly regional perspective; that is, it aims to locate its analyses within the South Asian region. This is done for two reasons. First, it seems important and potentially fruitful to identify gender systems as they operate within one major world region, with one unique complex of cultures. This should help to clarify and focus some of the issues raised, and often left hanging, in a world-systems approach to gender, or alternatively, in an internationally comparative approach to women and development (Young's introduction to Young, 1986, is an excellent summary of the issues).

Second, a specific contribution to modern South Asian studies is envisioned—one which can demonstrate the essentiality of gender analysis to the whole field, rather than keeping it removed from the mainstream in a women's studies category. Gender analysis ought to be seen as essential to the field for the following overarching reason. Studies of gender and political economy cannot be done as studies of women in isolation; they are, more broadly, *systems studies* which see gender as integral to systems of social relationships.

Two basic questions link gender to system. First, how do the different conditions and roles of women and men help us to better understand the workings of the wider system? And second, what role does gender, as constructed within the system, play in maintaining other important aspects of the system, such as property and class relations? All the contributions in this volume ultimately attempt to address some of the implications of these questions.

Political economy and gender are integrally interrelated, since we cannot fully understand gender until we find its roots in the structure of relations of human beings to the material base, which makes the continuation of society possible. What is more to the point at this juncture, we do not fully understand political economy—that nexus of relations of society to the material base—without incorporating genders as part of it.

To uncover the dynamic of the relationship between political economy and gender, an historical perspective, broadly defined, is needed. The historical conditions of daily life for women and men are important to understand, not only in themselves, but also to help us generalize about the emerging place of the sexes within larger systems, and the nature of those systems themselves.

SUMMARY OF THE CHAPTERS IN THIS VOLUME

The papers in this volume focus on historical and contemporary issues in the lives of women, within the specific regional context of South Asian societies and cultures. They present South Asian data either newly collected or freshly analysed, and explore theories of gender relations in a South Asian context. The chapters are grouped into three topic areas which seem of fundamental interest in elucidating the essentiality of gender to modern political economies.

INTRODUCTION

The papers do not represent all the subregions of South Asia. Neither is every nation of the region covered (Pakistan and Nepal gained no contributors for this effort), nor are those subregions which are covered necessarily representative of all of South Asia. This kind of enquiry cannot be exhausted in a single volume. The volume is, then, representative of a starting point for further efforts to link gender to political economic conditions and processes within the framework of the broad cultural-historical region of South Asia. There need to be more studies following the theme of this book.

1. Relations of Production/Reproduction

The processes that lead to the reproduction of life and society are basic to the construction of every gender system in every society. In South Asia, biological reproductivity has been, in comparative context, a defining characteristic of female roles to an extreme degree. But the relations of reproduction are not the same across regions, time periods, and social classes. They are closely linked to the state of development of the relations of production. This is illustrated in the chapters in this section.

Miriam Sharma's paper represents part of a new development in anthropology—the much-needed perspective of a feminist ethnography. She sheds light on the lives of women in their reproductive roles, carried out under multiple constraints, in village Rajasthan, and on the basis of a close reading of their own statements, pushes forward the theoretical boundaries of our understanding. The Jefferys' chapter is an ethnographic and micro-level view of reproductive women within a sociological framework. They continue their musing (following their 1989 book) about the larger ramifications of the relations of reproduction in a village in Bijnor District, UP. My own paper attempts to define some of the territory for understanding the historical and social demography of reproductive relations in India. I look at differences in demographic regimes in three of India's major regions between 1880 and

1930, and discuss what these differences may mean for Indian history, in the light of the concept of social reproduction.

2. *Class, Gender, and the Intensification of Economic Forces*

The fact that the categories of class, gender, and caste and/or race and/or religion cut across each other has been widely discussed by feminist scholars. A South Asian regional perspective is particularly useful in displaying the way in which these interrelations are locally unique in their particular configuration within this region. It also shows, however, that there are commonalities across locales which reveal more general, world-system dynamics. One of these dynamics is a tightening of the screws, a growing pressure to participate in wage-labour markets under disadvantageous working conditions, which is affecting women across a growing share of the globe. South Asian economies are no exception, whether the nation in question is explicitly open to the world market or not.

In two studies from India and one from Bangladesh this is demonstrated. Kalpana Bardhan explicates the socio-economic dynamic which maintains the class/gender system in India, and describes what this process is doing to women's lives as economic change occurs. She shows how calls for female solidarity miss the mark when they fail to view with understanding the stark realities of division which the cross-cutting categories of class, gender and caste/ethnicity impose on Indian women, but how a new kind of solidarity can emerge from this understanding. Priti Ramamurthy's study of rural Andhra Pradesh shows the way that modernizing forces in agriculture—changing irrigation technology, Green Revolution seed varieties, and credit-intensive forms of cultivation—lead to what might be called the mutation of modernization for women, as men find ways to claim the advantages. Women's increasing need to work for wages occurs under conditions whereby their relative status is weakened and made more vulnerable.

Shelley Feldman gives us another kind of view of the mutation of modernization for women. Into the urban, export processing zones of Bangladesh, impoverished educated women are drawn from rural areas, under the patronage of relatives of their village strongmen: and these are cultural conditions which are designed to prevent the experience from increasing the personal autonomy and power of the women.

3. *Gender in-Itself and for-Itself: Issues of Solidarity and Political Action*

Within the arena of political action we must also consider inaction: constraints on action. Here the critique of cultural consciousness has a rich potential and precedent for dialogue with political economy. The actions people take are vitally linked to their consciousness and ideology, as they attempt to live their beliefs and values. We may conclude that these are rooted in a false consciousness (this is implied in Caplan, 1985); or we may seek to find the meaning in them. This is undertaken by the first paper in this section. Jana Everett and Mira Savara examine a particular locus of action—organizations of informal sector workers in Bombay—and find a myriad of reasons why the women in these organizations struggle hard for ends that are mostly not their own. They analyse how effective five such organizations are in representing women and recommend specific measures for improvement.

The last two papers deal explicitly with issues of political mobilization. Amita Shastri illuminates the problem for women of having been granted rights without struggle, using the example of Sri Lanka; such rights can be and have been eroded once economic relations shift. Florence McCarthy's paper is about the conscientizing effects for rural Bangladesh women of involvement in development programmes—effects quite unintended, even undesired, by the agencies involved. This unintended conscientization is then followed by intentional struggle, as women mobilize around their own aims and needs.

Interesting commonalities and contrasts across regions within South Asia are observable from these papers. The long-term historical dynamics of gender system change in the context of colonialism and/or anticolonialism are discussed not only in the explicitly historical chapter (Clark), but also in several others (Shastri, Ramamurthy, McCarthy). In support of an historical perspective, we also find discussion of time-lagged social change elements in contemporary context (Bardhan, Feldman).

The most important of these, strikingly comparable across regions, is education. Educated women are being drawn into labour markets that do not require their education across the various countries of the region, particularly those with specific policies of world market integration, Bangladesh and Sri Lanka (Feldman, Shastri). Educated women's relation to the labour market across several countries shows the growth of middle-educated womanpower and of relevant labour markets, modified by differing political circumstances. These levels of education and employment are slow to promote the sort of quantum change in female autonomy that modernization theorists would suggest. On the contrary, uneducated women's mobilization in some cases is more sudden, even startling (McCarthy). In others it remains embryonic or just developing (Everett and Savara, Bardhan).

Demographic factors are widely discussed, not only those papers explicitly dealing with women's reproductivity (Sharma, the Jefferys, Clark), but in a large number of the others (Bardhan, Ramamurthy, Feldman, Shastri). This is highly appropriate in view of the essential need for understanding the demographic implications of the broader social systems and processes that have emerged in the demographic hotbed of South Asia. Gender and demography are inextricably linked and together need much more careful discussion within South Asian studies.

Here, demographic issues such as age structure, population density, population growth and urbanization are considered in several papers. Their effects on gender relations and class structure and the link between the two are examined. The role of fertility in structuring social relations, and vice versa, is richly considered in this volume. More empirical research in particular localities is needed on the crucial relationship between class and fertility, using a concept of class which is appropriate, rather than either a reification of abstract categories under which large-scale data sets are collected, or an uncontextualized measure of socio-economic level alone (as discussed in Clark, 1977). The theoretical proposals offered in these papers can be used as a basis for new hypotheses on class-fertility relations, in the form of a commentary on those that have been generated by demographers such as Caldwell, Cain, Das Gupta and Dyson (see references listed in Sharma's chapter and discussed in a note there).

Also discussed in a related context are the ecological effects of local population expansion, which can lead to lowered productivity, changing production relations and the tightening of the screws (Clark, Bardhan, Ramamurthy). This chain of effects is then linked to the increased entry of educated women into very low-skill jobs (Shastri, Feldman), in a kind of feedback loop that disadvantages women in new contexts (see also Agarwal, 1988; Beneria, 1982; Sen and Grown, 1987).

The distribution of health, and the particular vulnerability of women to ill health under the inequitable conditions imposed on them, is another theme found in several papers (Sharma, the Jefferys, Clark, Bardhan, Shastri). Health conditions, particularly those of reproductive health, are shown to be poor for women, not only in village areas that are not well integrated into intensive market conditions, but also in some of those that are. And if health status is low in the female wage-labour stratum in India, it is now being threatened, though initially higher, among the same in Sri Lanka, as economic and political conditions there become more difficult.

On a more theoretical level, several papers wrestle with the classic challenge (initially spelled out by Hartmann, 1979) of discussing the interrelationships between class and the other hierarchies that define women's statuses and life conditions, and try to overcome arbitrary categories and assumptions that do not fit well with South Asian realities, or the realities of women's lives in general (Sharma, the Jefferys, Clark, Bardhan, Feldman).

The nexus of kinship (early placed in feminist context by Rubin, 1975) is analysed in its relationship to the resources and autonomy available to women—or withheld from their grasp (Sharma, the Jefferys, Ramamurthy, Feldman). And, the emergence of new modes of interaction, new possibilities of association and conscientization, is explicitly examined (McCarthy, Bardhan, Everett and Savara) or implicitly considered in all of the papers.

PROSPECTUS FOR SOUTH ASIAN GENDER STUDIES IN THE FUTURE: THE LARGER PROJECT

Ultimately, we need to develop a more comprehensive perspective—across more of the various regions of the subcontinent—on the historically changing situation of women in relation to men within the various overlapping systems in which life and behaviour take place. This volume marks only a beginning to such a project. We need to cover more regions and countries and develop a broadly comparative framework regarding essential themes.

A list of relationships between gender and various systems that need more study and reflection, with specific focus on parts and locales of the South Asian region, suggests itself (and is by no means exhaustive):

gender and family/kinship
gender and the household economy
gender and the division of labour
gender and class formation
gender and the labour process/the mode of production
gender and educational opportunity
gender and the state of/the law
gender and ecology/resource exploitation
gender and population dynamics

gender and the capital accumulation process
gender and political mobilization

More studies are needed, then, on the sub-regional relations between the sexes, over time and across varying conditions, in marriage, kinship, childbearing and the family; in property relations, class structure, the labour force, and the informal sector; in politics, social activism, and the law.

Comparisons over time and space provide useful insights about differences among the conditions of South Asian women; and by implication about the conditions of women and men more broadly. How have life conditions differed across regions over time? How have women and men from different regions and circumstances fitted differently into wider political and economic systems? What have been the effects of colonialism and its consequences in shaping lives, by gender? What are the linked effects of these historical and social factors on the population situation? What about the intergenerational issues involved with gender: wealth and welfare, life and death, for the elderly and for children? How have these changed, under what historical and structural conditions?

The broader problematic for this kind of continuing inquiry is to develop a more fully adequate historical framework in which to understand the emergence of modern conditions of life. This involves a commitment to a material grounding along with a critical and sympathetic ear for cultural meaning. It involves also a specific regional focus within a wider world framework. The socio-economic conditions and dynamics of the South Asian region are rooted in that region's own cultural complex, unfolding within its own history—which is, however, a part of world history. The lives of men and women, gender relations, and all the development conundrums of the present and future are interconnected within this historically rooted yet ever changing web.

BIBLIOGRAPHY

Joan Acker, 'Class, Gender, and the Relations of Distribution', *Signs*, 13(3), 473-97, 1988.

Haleh Afshar, ed., *Women, Work, and Ideology in the Third World*, London: Tavistock, 1985.

Bina Agarwal, ed., *Structures of Patriarchy: State, Community and Household in Modernizing Asia*, New Delhi: Kali for Women, 1988.

Pranab K. Bardhan, *Land, Labor, and Rural Poverty: Essays in Development Economics*, New Delhi, 1984.

Pranab Bardhan and T. N. Srinivasan, ed., *Rural Poverty in South Asia*, New York: Columbia University Press, 1988.

C. A. Bayly, *Indian Society and the Making of the British Empire*, The New Cambridge History of India, Cambridge: C.U.P., 1988.

Lourdes Beneria, ed., *Women and Development: The Sexual Division of Labor in Rural Societies*, New York: Praeger, 1982.

Lynn Bennett, *Dangerous Wives and Sacred Sisters: Social and Symbolic Roles of High-Caste Women in Nepal*, New York: Columbia University Press, 1983.

Hans P. Binswanger and Mark R. Rosenweig, eds., *Contractual Arrangements, Employment, and Wages in Rural Labor Markets in Asia*, New Haven: Yale University Press, 1984.

Jeanne Bisilliat and Michele Fieloux, *Women of the Third World: Work and Daily Life*, London and Toronto: Associated University Presses, 1987.

James W. Bjorkman, ed., *The Changing Division of Labor in South Asia: Women and Men in India's Society, Economy, and Politics*, New Delhi: Manohar, 1987.

Ester Boserup, *Women's Role in Economic Development*, London: George Allen & Unwin, 1970.

Mead Cain, 'Women's Status and Fertility in Developing Countries: Son Preference and Economic Security', World Bank Staff Working Paper Series No. 682, Washington, D. C., World Bank, 1984.

Patricia Caplan, *Class and Gender in India: Women and their Organizations in a South Indian City*, London: Tavistock, 1985.

Patricia Caplan and Janet Bujra, *Women United, Women Divided: Cross-Cultural Perspectives of Female Solidarity*, London: Tavistock, 1978.

R. H. Cassen, *India: Population, Economy, Society*, New York: Holmes and Meier Publishers, Inc., 1978.

Karuna Chanana, ed., *Socialization, Education and Women: Explorations in Gender Identity*, New Delhi: Orient Longman, 1988.

Neil Charlesworth, *British Rule and the Indian Economy 1800-1914*, London: Macmillan, 1982.

Lincoln Chen, Emdadul Haq and Stan d'Souza, 'Sex Bias in the Family Allocation of Food and Health Care in Rural Bangladesh', *Population and Development Review*, 7(1), 1981.

Alice W. Clark. 'Mortality, Fertility, and the Status of Women in India, 1881-1931', in Tim Dyson, ed., *India's Historical Demography: Studies in Famine, Disease and Society*, London: Curzon Press, 1989.

―― 'Social Demography of Excess Female Mortality in India: New Directions', in *Economic and Political Weekly*, 22(17) Review of Women Studies, 25 April 1987.

―― 'Limitations on Female Life Chances in Rural Central Gujarat', *Indian Economic and Social History Review*, 20(1) January-March 1983.

Jane Fishburne Collier and Sylvia Junko Yanagisako, ed., *Gender and Kinship: Essays Toward a Unified Analysis*, Stanford: Stanford University Press, 1987.

Committee on the Status of Women in India, *Towards Equality*, New Delhi: Ministry of Education and Social Welfare, 1974.

Veena Das, 'Gender Studies, Cross-Cultural Comparison and the Colonial Organization of Knowledge', *Berkshire Review*, 1986.

Monica Das Gupta, 'Selective Discrimination against Female Children in India', *Population and Development Review*, 13(1) 1987.

Christine Delphy, *Close to Home: A Materialist Analysis of Women's Oppression*, London: Hutchinson, 1984.

Neera Desai and Maithreyi Krishnaraj, *Women and Society in India*, Delhi: Ajanta Publications, 1987.

Neera Desai and Vibhuti Patel, *Indian Women: Change and Challenge in the International Decade 1975-78*, Bombay: Popular Prakashan, 1985.

Ruth B. Dixon, *Rural Women at Work: Strategies for Development in South Asia*, Baltimore: Johns Hopkins University Press, 1978.

Tim Dyson and Mick Moore, 'On Kinship Structure, Female

Autonomy and Demographic Behaviour in India', *Population and Development Review*, 9, 1983.

Paula England and George Farkas, *Households, Employment and Gender: A Social, Economic and Demographic View*, New York: Aldine, 1986.

Jana Everett, *Women and Social Change in India*, New Delhi: Heritage, 1981.

Rehana Ghadially, ed., *Women in Indian Society: A Reader*, New Delhi: Sage Publications, 1988.

Paul Greenough, *Prosperity and Misery in Modern Bengal: The Famine of 1943-44*, Oxford: O.U.P., 1982.

Leela Gulati, *Profiles in Female Poverty*, London: Pergamon Press, 1982.

Barbara Harriss, 'Differential Female Mortality and Health Care in South Asia', *The Journal of Social Studies*, 44, 1989.

Heidi Hartmann, 'The Unhappy Marriage of Marxism and Feminism: Towards a More Progressive Union', *Capital and Class*, 8, 1979.

Charles Heimsath, *Indian Nationalism and Hindu Social Reform*, Princeton: Princeton University Press, 1964.

Renee Hirschon, ed., *Women and Property—Women as Property*, London: Croom Helm, 1984.

Ronald B. Inden and Ralph W. Nicholas, *Kinship in Bengali Culture*, Chicago: University of Chicago Press, 1977.

Doranne Jacobson and Susan Wadley, *Women in India: Two Perspectives*, New Delhi: Manohar, 1977.

Rounaq Jahan and Hanna Papanek, ed., *Women and Development: Perspectives from South and Southeast Asia*, Dacca: Bangladesh Institute of Law and International Affairs, 1979.

Devaki Jain and Nirmala Banerjee, *The Tyranny of the Household: Investigative Essays on Women's Work*, Delhi: Shakti Books, 1985.

Kumari Jayawardena, *Feminism and Nationalism in the Third World*, London: Zed Books, 1986.

Patricia Jeffery et al., *Labour Pains and Labour Power: Women and Childbearing in India*, London: Zed Books, 1989.

Promila Kapur, *Marriage and Working Women in India*, New Delhi: Vikas, 1982.

Irawati Karve, *Kinship Organization in India*, Bombay: Asia Publishing House, 1965.

Madhu Kishwar and Ruth Vanita, ed., *In Search of Answers: Indian Women's Voices from Manushi*, London: Zed Books, 1984.

Pauline Kolenda, *Caste in Contemporary India: Beyond Organic Solidarity*, Menlo Park, CA: Benjamin Cummings, 1978.

J. Krishnamurty, ed., *Women in Colonial India: Survival, Work and the State*, Delhi: Oxford University Press, 1989.

Annette Kuhn and Ann Marie Wolpe, ed., *Feminism and Materialism: Women and Modes of Production*, London: Routledge, 1978.

Joyce Lebra, Joy Paulson and Jana Everett, ed., *Women and Work in India: Continuity and Change*, New Delhi: Promilla, 1984.

Joanna Liddle and Rama Joshi, *Daughters of Independence: Gender, Caste and Class in India*, London: Zed Books, 1986.

K. Mahadevan, ed., *Women and Population Dynamics: Perspectives from Asian Countries*, New Delhi: Sage Publications, 1989.

David Mandelbaum, *Human Fertility in India: Social Components and Policy Perspectives*, Berkeley, 1974.

Karen Oppenheim Mason, *The Status of Women, Fertility and Mortality: A Review of Interrelationships*, New York: Rockefeller Foundation, 1984.

Joan Mencher, *Agriculture and Social Structure in Tamil Nadu: Past Origins, Present Transformations and Future Prospects*, Durham, NC: Carolina Academic Press, 1978.

Maria Mies, *Indian Women and Patriarchy: Conflicts and Dilemmas of Students and Working Women*, New Delhi: Concept Publishing Co., 1980.

——, *Patriarchy and Accumulation on a World Scale*, London: Zed Books, 1986.

Barbara D. Miller, *The Endangered Sex: Neglect of Female Children in Rural North India*, Cornell, 1981.

——, 'Daughter Neglect, Women's Work, and Marriage: Pakistan and Bangladesh Compared', *Medical Anthropology* 8(2) 1984.

Asok Mitra, *India's Population: Aspects of Quality and Control*, New Delhi: Abhinav Publications, 1978.

Rosalind O'Hanlon, 'Recovering the Subject: *Subaltern Studies* and Histories of Resistance in Colonial South Asia', *Modern Asian Studies*, 22(1) 1988.

Gail Omvedt, *We Will Smash This Prison! Indian Women in Struggle*, London: Zed Books, 1980.

———, *Women in Popular Movements: India and Thailand during the Decade of Women*, Geneva: United Nations Research Institute for Social Development, 1986.

Hanna Papanek and Gail Minault, ed., *Separate Worlds: Studies of Purdah in South Asia*, Delhi: Chanakya Publications, 1982.

Nenneke Redclift, 'Rights in Women: Kinship, Culture, and Materialism', in *Engels Revisited*, Janet Sayers et al., London: Tavistock, 1987.

Nenneke Redclift and Enzo Mingione, ed., *Beyond Employment: Household, Gender and Subsistence*, Oxford: Basil Blackwell, 1985.

Mark R. Rosenzweig and T. Paul Schultz, 'Market Opportunities, Genetic Endowments, and Intrafamily Resource Distribution: Child Survival in Rural India', *American Economic Review*, 72(4) 1982.

Gayle Rubin, 'The Traffic in Women: Notes on the "Political Economy" of Sex', in *Toward an Anthropology of Women*, Rayna Reiter, ed., New York: Monthly Review Press, 1975.

Kumkum Sangari and Sudesh Vaid, ed., *Recasting Women: Essays in Colonial History*, New Delhi: Kali for Women, 1989.

Joni Seager and Ann Olson, *Women in the World: An International Atlas*, New York: Simon and Schuster, 1986.

David M. Schneider and Kathleen Gough, ed., *Matrilineal Kinship*, Berkeley: University of California Press, 1961.

Amartya Sen, 'How is India Doing?', in D. K. Basu and Richard Sisson, ed., *Social and Economic Development in India: A Reassessment*, Delhi: Sage Publications, 1986.

Amartya Sen, *Resources, Values and Development*, Oxford: Basil Blackwell, 1984.

Gita Sen and Caren Grown, *Development, Crises and Alternative Visions: Third World Women's Perspectives*, New York: Monthly Review Press, 1987.

Miriam Sharma, 'Caste, Class and Gender: Women's Role in Agricultural Production in India', *Journal of Peasant Studies*, 12(5) 1985.

Ursula Sharma, *Women, Work and Property in North-West India*, London: Tavistock, 1980.

Andrea Menefee Singh and A. Kelles-Viitanen, ed., *Invisible Hands: Women and Home-Based Production*, New Delhi: Sage, 1987.

Joan Smith *et al.*, ed., *Households and the World Economy*, Beverly Hills: Sage Publications, 1984.

—— ed., *Racism, Sexism, and the World-System*, New York: Greenwood Press, 1988.

M. N. Srinivas, *Some Reflections on Dowry*, Delhi: Oxford University Press, 1986.

Pravin Visaria, *The Sex Ratio of the Population of India*, Census of India 1961, Monograph 10, New Delhi: Office of the Registrar General, 1969.

Lise Vogel, *Marxism and the Oppression of Women: Toward a Unitary Theory*, New Brunswick: Rutgers University Press, 1984.

Susan Wadley, ed., *The Powers of Tamil Women*, Syracuse: Maxwell School, 1980.

Kathryn B. Ward, *Women in the World-System: Its Impact on Status and Fertility*, New York: Praeger, 1984.

Stanley A. Wolpert, *Tilak and Gokhale: Revolution and Reform in the Making of Modern India*, Berkeley: University of California Press, 1962.

Women: A World Report—A New Internationalist Book, New York: Oxford University Press, 1985.

Kate Young, Carol Wolkowitz and Roslyn McCullagh, ed., *Of Marriage and the Market: Women's Subordination in International Perspective*, London: CSE Books, 1981.

Kate Young, ed., *Women and Economic Development: Local, Regional and National Planning Strategies*, Oxford: Berg/UNESCO, 1986.

2

Engendering Reproduction:

The Political Economy of Reproductive Activities in a Rajasthan Village

MIRIAM SHARMA AND URMILA VANJANI

I. INTRODUCTION — REPRODUCTIVE POLITICS

Marx first suggested looking for the dialectical relationship between the natural and social side of reproduction (see Sharma 1985). These exist not as parallel and independent processes, but as closely entwined and interrelated. In any discussion of reproductive freedom the central issue becomes that of the 'social and material conditions under which choices are made' (Petchesky, 1979: 105; also Gordon, 1977: 408). Reproduction under constraints of definite material conditions set the limits to the natural process (Petchesky, 1979: 105). While reproductive freedom—the ability of women to control their own fertility and its consequences—is an aspiration, the right to choose (or to maximize, in neo-classical parlance) means little for those who are poor and powerless. Women, in such situations, may resort to abortion or infanticide with impunity, but that act of 'resistance' grows out of female powerlessness and subordination (Petchesky, 1979: 98, 105, on abortion). This is most noticeable in the case of female infanticide, which may be seen as complicity with the norms and values of patriarchy.

The methods, goals, and control of women's reproductive activities become contested terrain, particularly between men and women, within the household arena. Recent scholarship on women has made a convincing case of locating the roots of female oppression within the domestic sphere. The source of the subordination of women lies within the parameters of their households and in the social relations of gender. This understanding has added a new dimension to traditional analyses, including the Marxist.

In India—and especially in the north—the inadequate recognition and low evaluation of women's role in the household is a major factor contributing to their low status and welfare (i.e. their subordination). The situation is as Kate Young *et al.* describe in their introduction to *Of Marriage and the Market*:

While the household is neither a natural unit nor a universal category, the domestic sphere, its separation from the public or social world, and relations between men and women within different households, are fundamental to gender subordination.... Family based households... are hierarchical structures characterised by male dominance. Women are made vulnerable in the isolation of domestic life, and are largely dependent on the men who represent them 'outside'. Women generally, and wives in particular, often do not have direct access to the market .... (1988: ix).[1]

Most women's work is not destined for the market and hence their economic contribution is largely ignored (Jain and Banerjee, 1985: xvi, xviii). Further, more and more women are seeking employment of an unskilled type involving hard manual labour (Mukhopadhyay, 1984: 28).[2] Intra-household relations, however, must be viewed within the context of the wider society of which they are part and the sexual division of labour in that society. (Petchesky, 1979: 104; Young *et al.*, 1981: x). Minimally defined as 'the system of allocating particular tasks to men and others to women' (Young *et al.*, 1981: x), this division gives little value to the domestic labour and reproductive work that falls to women. The sexual division of labour expresses, embodies and perpetuates the gender subordination embedded in it (Mackintosh, 1981: 2–3).

Into this wider arena, as Petchesky astutely notes, comes the 'conjuncture of medical, corporate, and state interests on the "management" of reproduction (that) has defined the choices of all women, but in a way that is crucially different depending on one's class and race' (1979: 106). In Third World countries, as in the other worlds, the organised forces that have emerged to shape the class-specific socially constructed character of women's reproductive experience are powerful and diverse (Petchesky, 1979: 106). The role of the state has assumed critical proportions where population control policies have been adopted almost uniformly as key elements in 'development' strategies.

Within India, family planning has become a euphemism for sterilization. The government sees the programme as 'a remedy for almost every social ill' (Desai and Krishnaraj, 1987: 236). Underlining the rationale of all such policies is the assumption that population growth is the major determinant of the impoverishment of the rural poor. While some voices question this reasoning (Mamdani, 1972; Desai and Krishnaraj, 1987: 235ff; Epstein and Jackson, 1975: 6ff), they are not heard (ignored?) in government and policymaking circles.[3] More poor people certainly means more poverty, but that is not to be confused with its cause. A reduction in numbers of the poor will hardly promote better distribution of resources as long as structures of gender and class inequality remain (namely the vast literature on the 'Green Revolution'; Sharma and Vanjani, 1989 and other citations on the 'White Revolution').

While birth control methods for men (condoms and vasectomies) are far safer and simpler—causing less discomfort and risk—India's population control policy targets women primarily, it may be fair to say, almost exclusively. It is interesting to note that a leading cause of Indira Gandhi's downfall in the north (especially Uttar Pradesh) was the outrage prompted by her attempt to enforce vasectomies during the Emergency Rule in 1976. There appear to be no such difficulties in pursuing female sterilization. In this way, the state and family planning measures combine with patriarchal control in the family to reinforce the subordinate position of women and the manner in which men control them. It takes advantage of a cultural situation where 'men will not take responsibility for limiting the size of their families' (Desai and Krishnaraj, 1987: 238). The state perpetuates class and gender inequalities through policies which fall most heavily upon the poor and upon poor women most of all. Governmental and private population control agencies join with the intervention of the medical profession as 'medical indications' and 'medical effectiveness' become 'euphemisms of technical efficiency in population control' (Petchesky, 1979: 106). Large-scale commercialization of birth control products and services has meant that multinational pharmaceutical companies have vested interests in controlling reproduction and are important influences on the kinds of methods available to women, their safety, and their health.

It now becomes quite clear that women's status is a critical determinant of the manner in which they are integrated into the reproductive policies of the state. As Ann Stoler noted while speaking about Java (1977), and Naila Kabeer expanded upon in connection with women in Bangladesh (1985), status is concerned with two aspects—social power and autonomy. The former reveals the 'extent to which they have command over the social product', while the latter relates to 'the ability to control various aspects of personal life' (Kabeer, 1985: 86). The analysis of unequal gender relations specifies the ideological and structural constraints that inhibit women's enjoyment of power and autonomy. In turn, this status is linked to their low position within the household. The circle is complete, if not completely closed.

It becomes equally apparent that the contradictions inherent in the problematic of unpacking gender and reproduction are manifold: women's work makes the most contribution to the household—yet it is the least valued; women have the greatest nutritional needs (in terms of work output and childbearing)—yet they have the lowest nutritional intake (see especially Batliwala, 1985); women as childbearers are critical for population survival (i.e. one man may impregnate many women)—yet there has been a steady decline in the sex ratio since the beginning of the century;[4] poor families need high fertility because of high infant mortality and the unstable circumstances in which they must survive—yet poor women lack adequate food, water, economic security, health facilities, and suffer further ill health due to multiple pregnancies in a weakened condition; women are valued most for, and defined by, their childbearing role—yet the needs associated with this role are given little weight (Jeffery et al., 1989: 221).

This discussion approaches these contradictions by attempting to understand the knowledge of, and control over, reproductive activities—menstruation, sexuality in marriage, pregnancy, childbirth, and family planning—for a specific group of women. The analysis proceeds by a discussion of this knowledge and the actual experiences of women in the village of Shankpur in Alwar district, Rajasthan, some ninety miles from Delhi. This is an overwhelmingly rural area that borders on the neighbouring state of Haryana and is in close proximity to the nearby state of Panjab. For eight months we lived in intimate contact with the village women, and especially with a group of poor women who qualified for a government loan to purchase a milch animal. We also had a series of in-depth interviews with a number of women about their reproductive histories and experience.[5]

The next section looks at women's work and health in the context of the household. The subsequent sections deal with what women told us of their knowledge of reproductive activities and actual experiences with menstruation, sexual relations, pregnancy, childbirth, and family planning. The final section presents some of the conclusions that link an analysis of the data to the key ideas discussed in the introduction.

II. WOMEN'S WORK AND HEALTH IN SHANKPUR VILLAGE[6]

The village of Shankpur represents a fascinating mix of ethnic groups, castes, and religions. Almost half (44 per cent) of the 1800 inhabitants of Shankpur are Panjabi Jats—who had migrated to the area at the time of Partition in 1947. Originally coming from the fertile land of five rivers in Multan (now in Pakistan), Jats are known as sturdy peasants who span all across northern India from Panjab to Uttar Pradesh.[7] They are the agriculturalists who have created the 'green revolution' in Panjab. Their coming transformed the village from a predominantly Muslim (Meo) one into a Hindu area. The next largest group in the village are the Untouchables who are 25 per cent of the population; these include Chamars and Bhangis, the latter representing only five per cent. The Ahirs, traditional cow-herders, comprise 14 per cent of the village population today. They, like a small segment of the Panjabi Jats, live all together in a separate hamlet (*dhani*) near their fields. They own a proportionately larger amount of land than any other caste. Muslims represent 10 per cent of Shankpur people. These include the Meos[8] (some of whom live in their own dhani) and Fakirs. The three remaining castes in the village are Kumhars (6 per cent), Banias (1 per cent), and Nai (2 households out of a total of 252).

Shankpur represents a fairly 'egalitarian' village in terms of both status, landownership, and wealth. The great disparities visible in other villages and parts of India are not readily apparent. The reasons for this are twofold. On the one hand, the traditional pattern of land concentration in place before 1947—where the 'dominant' caste and premier landholders were the Meos (many being extremely large landlords)—was broken by their migration to Pakistan and the reallocation of this land in small plots to Panjabi refugees. Land ownership also became a reality for a number of Chamar families. On the other hand, the rather limited success of the 'green revolution' here, due to the scarcity of water, has retarded the acquisition and alienation of land that has become a hallmark of areas of highly capitalized agriculture.

One of the most striking things about a village in Rajasthan is the intensity and duration of women's work. A walk down the main street of Shankpur, at perhaps any time of the year except when mustard and wheat are being harvested in April and May, reveals a number of men playing cards in small groups. At any time of the year (except the two summer months)—between the hours of 8 a.m. to 4 p.m.—few women will be seen; it is definitely difficult to meet a woman or a teenage girl from a nuclear family, for they are in the fields. Girls are disproportionately kept from attending school, and it is not uncommon to see little girls of even 6 to 8 years assuming all the household and nurturing tasks while their mothers work in the fields.

Twice a day women go to the fields to cut, collect, and then carry huge headloads of fodder home, where the animals (buffaloes or cows) are kept tethered nearby or in the courtyard. Twice a day they will feed and milk the animals, also cleaning the shed or area where they are tied, and collecting the dung to make into fuel cakes. During the three winter months they make numerous six-hour long trips to the hills near the village to collect firewood, the main source of fuel. They will be at the wells bathing and washing their children, animals, and clothes. Twice a day, they will need to balance ceramic pots on their heads to carry water from the well to the home for drinking, cooking, and cleaning. If they are poor Meo, Panjabi, and Chamar women, they also will be working in someone's fields as *mazdoors* (labourers), earning 6 or 7 rupees a day. If they own land, they will be labouring in their fields at sowing, reaping, and irrigating times.

Here in Rajasthan, women do all the work that men do (except ploughing)[9] as a matter of course, while men do none of what is considered women's work.[10] It is not surprising that among the elderly Panjabi women, all uniformly bemoan the fate that brought them to this *bukkha mullak*—region of hunger—called Rajasthan. In their pre-Partition home in western Panjab women did little work outside the home other than harvesting the cotton crop, but their families could not have survived those early years in Shankpur if women did not do this work. A family cannot survive today without the work of the women. Both Panjabi men and women say that when they first came, they saw all the work that the local females were doing and they followed suit.

It was not surprising to learn, then, that overwork and lack of adequate nutrition are the two root causes for the majority of women's ailments. Some of the daily maladies which strike them, and about which they spoke often and sometimes came to us for help, are fever, headaches, upset stomachs, diarrhoea, exhaustion, general body and specific muscle pain, burns, cuts, boils and other skin infections, and irregular menstrual cycles: these present a starting list.[11] Uncorrected weak eyesight and loss of teeth take their toll quite early in a woman's life, as does the excessive bleeding, pain, and general weakness that actually occurs or is imagined by those who undergo a tubal ligation for the prevention of further pregnancies. 'Home-remedy' abortions, even more than medical ones, carry tremendous risks for the women. Severe bleeding, complications—even death—and lasting weakness and ill-health from multiple pregnancies are a reality for all Shankpur females. Adequate rest and nutritional care for the mother and child is not easily available and delivery occurs at home under unsanitary conditions, with only a midwife in attendance. Insufficient breast milk is also a frequent complaint and cause of great worry. Beyond this is the whole area of serious illness that occurs: tuberculosis, cancer, typhoid, rheumatic fever.

The distribution of good health in Shankpur, as elsewhere in India, has a definite class bias (see Zurbrigg, 1984). The poorest women are the most overworked. Untouchable Chamar and Bhangi women often labour under the triple burden of home/child care, *mazdoori* (wage labour in the fields), and animal care (if they have any). Landless women of other castes (except the Bania-merchants) are in the same bind. Overworked and already weakened by multiple pregnancies, these women have the poorest diets. Most commonly there is a morning 'meal' of dry *roti* (i.e. 'bread' without *ghee*, clarified butter) and perhaps some lentils, preceded by a cup of tea. The evening meal consists of roti or rice, more *dal* (lentils) or a seasonal vegetable (potatoes and mangoes are among the most popular). Hunger in-between meals is quenched by consuming leftover roti. With an already low caloric diet, these women also had the least access to medical care. There are three 'doctors'[12] who have medical stores in the village where they sit and daily dispense their pills and injections. Such medical care is not cheap—a few pills can cost two to five rupees and an injection five to ten. A female labourer earns six rupees for a full day's work. It is understandable that the women wait for a long time before going to these doctors. What often occurs is that the woman's case progresses until it is beyond the capability of the village doctors and she must be taken to a government hospital in a nearby bazaar-village, an hour's walk away.[13]

In assessing the health of rural Indian women, attention must also be paid—as it hardly ever is—to their mental health. Headaches and even body pains are good indicators of the frequency of stress, though other causes are, of course, acknowledged. Women have so many worries—daily family survival; would their husband or son find work, would the crops do well (and all the myriad of worries dealing with assuring an adequate crop and harvest); when would the *akal* (drought) end; how will arrangements for a son or daughter's marriage be made, how would the expenses be met, and others. Some women live under daily tyrannies—their husbands drink excessively and/or physically abuse them.[14] Men's liaisons, dalliances, and even 'marriages' with other women are not unknown. Most women suffering from this sort of mental stress appear to draw upon an enor- mous reservoir of inner strength to cope, which often, but not always, includes the support of other women and family members. Suicide was rare. Much more common were the verbal expressions of the women, 'I would die', or 'Let God take me now'. It is within this situation of life, work, and general well-being (or lack thereof), that the political economy of reproduction must be understood.

III. KNOWLEDGE AND CONTROL OVER REPRODUCTIVE ACTIVITIES

A. Menstruation

All the women told us that before their menarche they had known nothing about the coming of menstruation and what it actually means physiologically. At the most, a female relative other than the mother, or even a neighbour or girl friends, might have asked whether menstruation had begun—*kaparda aya*—literally 'cloth came', referring to the cloth used to absorb menstrual blood. Afterwards, they were only told that they had become mature (*syani ho gayi*), that this occurred to all women, and that now they could have children. It was usually the older brother's wife (*bhabhi*) who explained this much and instructed the young girl on how to wear underpants (*kaccha*) and keep some cloth to absorb the blood.[15] If the onset of menstruation occurred in their husband's house (*sasural*), it was the husband's older brother's wife (*jethani*) who spoke with them. Other information conveyed to them was that those who don't have their periods can 'fall sick and have diseases'.

The actual onset of menstruation (*mahina*) was remembered as a frightening experience by all of the women.

My mahina started just about a year before my marriage. I didn't know (understand) anything; I was afraid. I didn't tell (anyone at first) and managed myself. Sometimes bhabhis (older brothers' wives) tell girls to be careful. After I told my bhabhi, she explained about it.

It happened two years before marriage. I knew nothing. I was working and felt something; I started to cry and tell my bhabhi. All she said was, 'It happens to all women'.

My periods started in my in-laws house (sasural). I told my husband's older brother's wife (jethani), mother-in-law (*sas*), and husband's sister (*nanad*). They said '*kaparda ho gaya*' and they were laughing and laughing, but didn't explain anything to me. Then I washed and changed my clothes. But I didn't wear underwear. I didn't even use any cloth and let me petticoat get dirty. The whole day I was working and didn't know anything about it. Now girls wear underpants (kaccha) and keep some cloth underneath (there). But in our time there was no fashion of wearing kaccha. Nobody told me anything about it. Now I have learned and I started keeping some cloth when children were born.

I didn't know anything. When I saw a blood spot I told my friends. I didn't like it all (*bura laga*) and felt afraid (*dar laga*). I didn't tell my mother; my friends said that it happens to all women.

I was about 14 years. It was the day of my *gauna* (second marriage ceremony to mark going to live in the husband's home). I didn't know anything. My bhabhi used to ask me whether my mahina had started. I didn't understand what she was asking about. A lady neighbour also asked—but I didn't understand. It started the day after the gauna when I was leaving for my in-laws' house. My clothes got dirty. I was so afraid; I thought that whatever they had all been talking about had happened today. My in-laws saw what happened and gave me a petticoat to change into. But no one explained anything—my husband and I slept separately because of my mahina.

I was about 13 or 14 (four to five years before my marriage) when my mahina started. I knew nothing. I felt like my daughter Muni's age. My mother's brother's wife (*mami*), was newly-married then and I spent most of the time with her. But as she had just married she didn't know much; what she knew, we talked about. We were very close. When my mahina started, I told her and she said, 'It happens to all women'. She had asked me about it before (*kya kaparda ho gaya?*), but I didn't know what kaparda meant (because it just means cloth). I was in the fields when I went to urinate and saw blood. I became pale, so scared, and felt a pain in my stomach. I went to tell my mami—she said grown-up (*jawani*) girls don't mind about this. But I was so young, I didn't care (what she said). I kept on thinking, what has happened to me? I told my mami, not mother, that whatever you were talking about has happened. She gave me her underwear. She said you have become *syani* (mature); she didn't say anything else. Perhaps she also didn't know any more as she was still young and had not yet had a child. She told me how to keep some cloth in my underwear and not keep it lying about, but to put it where no one would know—not even my mother. One day my mother asked me whether my mahina had started—I lied and said no. As I grew older, I began to understand more.

In four of the cases, the women began to menstruate in their sasural after their second marriage ceremony (gauna) had taken place. In only one of these cases, however, did the woman sleep with her husband before that time. Women manage by wearing cotton underpants (kaccha) and stuffing some old cloth in them, which they were taught to 'not keep lying about', but to put 'where no one would see it—not even mother'. They were told to 'wear proper clothes so that nothing would be seen'. In the language of talking about menstruation, women used the words mahina (month) and kaparda (cloth), *dar gayi* (frightened) and *malum* (know) or *samajhana* (to understand) most often to express their experiences. Despite menstruation being regarded as causing a polluting state—when food should not be touched—all the women reported that they now cook during that time.

B. First Sexual Relations

From an early age, young girls are socialized so that they may learn to acquire and cultivate what is perhaps their most singularly important quality—that of *sharam*, best glossed generally as shyness or modesty, rather than shame.[16] A little girl soon learns to become modest about her decorum—the clothes she wears, her body movements, how she speaks—as well as where she goes and with whom she plays. Such training to acquire a quiet demeanour is ultimately geared towards the suppression of the increasing sexuality that comes with the onset of puberty. A critical symbol of this is the greater importance given to the *dupatta*, the long scarf that is to cover not only the girl's head, but also her growing breasts—the most public notice of her transformed sexuality.[17]

Jeffery *et al.* describe how the onset of puberty in a village in Uttar Pradesh is viewed as a time when children come to possess 'an immensely disruptive potential which must be carefully controlled and channeled.' They discuss perceptions of sexuality that are 'closely linked with a pair of opposites commonly used in local medical modes, *garmi* (heat, activity, stimulation) and *thand* (coldness, calmness, pacification)' (1989: 24). Foods are specifically linked to these two qualities, e.g. meat, fish and eggs which are usually taboo for women are *garam* (hot). Both semen and menstrual blood also are regarded as being garam and this puts both adolescent boys and girls in particularly susceptible states. The difficulties of regulating youthful sexuality, in a situation where 'the nagging worry for parents is that unbridled (the girl) will allow wrongful access to her body and bear the child of a man without formal rights over her' (1989: 25), leads to arranging early marriages.

In Shankpur—as in Rajasthan as a whole—a girl's marriage (*shadi*) often takes place before puberty but she usually does not go to live in her husband's home until that occurs, or even several years later, after a second ceremony (gauna).[18] Very young children may witness or hear their parents during sexual intercourse as they sleep with their mother, and even when grown, as completely private sleeping areas are not always available.[19] However, there is no extended discussion or explanation of what occurs during intercourse—especially for young girls. Kapur, in his study of sexuality among Kumaon villagers, also found that there 'is no systematic knowledge of sexuality' or of sexual technique (1987: 41, 54). Most particularly, even though considerable extramarital sex appears to be a regular feature of village life, 'conspicuously absent is intercourse between unmarried couples; it is between boys and men with older women' (1987: 40).

The women we talked with all explained that they were basically ignorant of what takes place between a man and a woman and viewed their 'wedding nights' as a rather traumatic experience.[20] Basically a girl is told by married friends in the village, by their sister-in-laws, bhabhis or—more generally jethanis—that she is expected to go along with her husband and to do what he wishes. Details of exactly what to expect were not provided (nor available) to any of the women and this became clear from their memories of the first experience with their husbands.[21] The other point that commands stark attention is the advice almost all recounted having received: be quiet, don't shout or scream.[22] One can only imagine the effect of such 'advice' on a young 15 or 16-year-old girl, socialized into a sublimation and innocence of her sexuality, as she encounters a stranger—her husband.

I don't have a jethani, so my nanad (husband's sister) came and said, 'Don't scream, just sleep nicely with your husband so people won't know (what's happening); otherwise everybody will find out.' It was very hard. I was small and knew nothing. For two to three visits to my sasural I was so afraid of my husband. But then I got used to it.

My bhabhi told about husband-wife relations. She said, 'Don't scream if your husband holds or touches you. Don't say to him "Why are you sleeping or doing like that".'

My friends talked about sex—those that got married first would talk and that's how we got to know about it. They would say that when you get married you will sleep with your husband—don't shout or scream.[23]

My jethani told me. She said they just make the girl sleep and when she is asleep others (sleeping in the room) would leave and then close the door. It was not explained what would happen. My bhabhi told me to keep quiet and not to scream.

I didn't know anything about sex. Even my husband did not know about it and we were both afraid.

After my gauna, my mahina started. One time I went to my *pihar* (natal home) and when I returned we slept together. Probably his brother told him and (it was) my sister (who) told me. She said, this is our custom (*riwaj*)—you don't need to cry; nobody should hear anything. She didn't explain anything.

As in the case with menstruation—but even more so—all the women recounted fearful experiences of their first intercourse:

I was so afraid in the beginning. I had one small room and as soon as my husband came, I hid. He said, 'Come out; why are you scared? Will I eat you?' Then he dragged me out by my hand. Then he taught me about it.

I was afraid—oh yes. In fact, I would not go to sleep and I would not let my jethani sleep either. Finally I slept and then she went away. Then my husband came and he woke me up. My jethani had explained what would happen. My husband didn't explain (samajhana) anything.

The first night, my nanad left me alone and then my husband came. He told me not to be afraid, but he didn't explain anything. He was very young (*jawan*) and I was small.

After marriage when I came to my sasural, my jethani was living there. After we finished eating, she went upstairs to the roof and spread my cot there and said 'you go sleep there'. But I was scared so I said no, I will sleep here (downstairs). She said, 'no, no, go and sleep on the roof.' But I didn't go and I slept outside my room on a cot. But I couldn't sleep because in my pihar all of us, brothers and sisters, used to sleep together. Here in an unknown house I was feeling alone. Then I went to urinate. It was dark, so I couldn't see my husband. But he saw me; he was upstairs. Then I took my *chadar* (bedsheet) and went to sleep inside the room. Nobody was there because everybody sleeps outside in summer. Afterwards, my husband came and said, 'Why are you sleeping on the floor? Come and sleep on the cot.' I said, 'No, I will sleep here.' Then he tried to hold my hand and I pushed it away. But my mami (mother's brother's wife) told me that a girl has to stay in her sasural after marriage and do whatever they ask. So, reluctantly, I finally went up on the roof and slept there. The next day I told my jethani that she did a very bad thing to me. She laughed and said from now on, every day, you have to sleep there. I said, no, I am going to my parents' house. My jethani was laughing and said—now this is your house.

I didn't know anything about sex. I was afraid the first time. I was sitting outside the room because I was afraid. My husband explained some things to me and then I was not afraid and I went to sleep. My sas was also sleeping there and so I thought she will stay with me. But when I fell asleep, my sas left and my husband came. My husband explained some things to me and then I was not afraid. He said, 'There is no need to be afraid.' Then he turned on the radio so no one could hear us. The next day, my sas and jethani also told me about it.

I slept with my husband even before my mahina—I was about 16 or 17 when they began. My friends told me (about it). They said, 'Don't shout, and keep quiet when you sleep with your husband.' I was afraid, but what could I do? My husband would beat me and make me sleep on the cot (*mar pitkar khat par sula deta tha*). I was crying and crying. Then my *sas-sasur* (in-laws) came and my husband got mad. Then he beat me again. I was more scared, but what could I do. I started sleeping with him slowly, slowly. These are the ways of the world (*sansar ki reet*), so what can I do? My sas-sasur put me in the room the first time. My sas said, 'Be quiet, otherwise your husband will be mad. Don't shout or say anything.' I was so small and didn't know anything and so I went quietly and slept. My husband did not beat me again. He explained—this is the *reet* (*riwaj*, custom); if the husband and wife don't sleep together then how will children be born. That's why girls and boys get married. The first time my husband hit me and then I became quiet and he explained. I didn't talk to him because I was shy. I hadn't seen this before and that's why I shouted the first time. Later I understood everything.

Several of the women said that they ran from their husbands in fear and then hid in some other part of the house to sleep. Although most mentioned some element of force, only a few spoke of being beaten on their wedding and subsequent nights. It is only her socialization into the supremacy of the husband's wishes in all matters—including the sexual—that permits the young bride the needed breathing space in which she may come to adjust to this aspect of her new life. And yet despite their own experiences, these women are not able to tell their own daughters anything. As one woman said, 'I will not explain anything to my daughter. Mothers don't tell their girls. Either her friends or some other relatives may tell the girls, but the mother does not.' The women are shy before their daughters '*beti se sharam* (or *sharm*) *ati hai*'. Further, as these same women come to bring a bride (*bahu*) into their own house, the cycle of concern for control over sexuality begins again. It would appear that this is less indicative of a 'complicity' in their own oppression than of the ways in which women are 'empowered' in different ways throughout their life cycle (see also the conclusion).

While neither the young man nor the woman are consulted about their choice of marriage partner, once married they play out their unequal relationship within the locus of the household. In every aspect of her daily life the bahu is controlled by others, and most specifically through her husband's control of her sexuality. In their analysis of the bahu in her sasural, in a chapter appropriately titled 'Someone Else's Property', Jeffery *et al.* continue their excellent discussion of a woman's sexuality and in whose control it lies.

A young woman's sharm (*sharam*, modesty) relates most obviously to her sexuality. Her transition to bahu necessarily entails sexual experience, but sexual intercourse is a *sharm-ki-bat* (shameful matter) not to be alluded to or publicized. Further, sexual intercourse is polluting. While a man can cleanse himself by washing, a woman is permanently affected by her conversion from virgin to wife . . . . A bahu and her husband should give no hint of the sexual dealings between them. She should keep her head covered in his presence and maintain a cool distance from him. Conjugal sexual activity is expected and enjoined: but a bahu's demeanour should not imply that she is sexually active. While the young couple share a *chula* (cooking place) with his parents, his mother may regulate when they sleep together (at least at home at night). Otherwise, however, the husband has the right of sexual access to his wife at times when he decrees: she should neither take the initiative nor refuse him. Many men . . . were explicit that sexual intercourse involves the exercise of power over a woman who is degraded in the process. An essential component of a husband's rule is sexual power over his wife (1989: 29).

Control of women's sexuality is, of course, not peculiar to India or the third world. In an insightful article, 'Sexuality and Control of Procreation', Mirjana Morokvasic points out how sexuality based on the double standard in Yugoslavia resembles

a contest in which, if sexual intercourse takes place, the man is considered the winner, the woman the loser. The woman can only win if she refuses successfully: in that way she not only keeps her reputation but becomes even more desirable. One of her best weapons is fear of pregnancy—in fact it is a fear for her and a threat for him. *The technology of modern contraceptives deprives women of that tool.* Another weapon is pregnancy itself, though this has a somewhat different purpose (1981: 137, my emphasis).

While Morokvasic's work reveals some fascinating complexities and contradictions in the relationship between contraception and abortion—between sexuality and procreation—which will be discussed further, it is her comments on sex and power that are of relevance here. The semiotics of this inequality between the woman and the others who control her within the village household are expressed in myriad ways. In her general dealings, a woman defers to her husband and should be slow to offer an opinion or to criticize. She remains invisible and voiceless before men elder to her husband and subservient to her mother-in-law and elder sister-in-law. Although responsible for household food preparation, she eats the last, unseen by others. In physical movements she must be modest and act with restraint, sitting on the floor—not the cot or a chair—before older males; her mobility is severely restricted and controlled. Lack of education also serves to put constraints around the mental mobility of a woman. Often illiterate, her knowledge about the world around her is also limited. The greater mental and physical mobility afforded to boys accounts for the 'knowledge gap' regarding reproductive activities apparent between them. Although this picture is painted in broad strokes and admits significant individual variation, it is particularly valid for new brides.

A woman comes to wield power in the household through her procreative activities. Yet here too, the structural patterns leading to her subordinate status repeat themselves. A woman initially has no prior knowledge of what childbirth entails and her experiences with procreation and repeated pregnancies ultimately serve to weaken her physically.

IV. EXPERIENCES OF CHILDBIRTH

There are a number of factors that have an impact upon the actual number of pregnancies, deliveries, births, and infant mortalities that become part of the life-history of individual women. In hearing women's stories these events become not just statistics, but an intimate matter of experience and survival.

Under conditions of poor nutrition and ill-health that plague vast numbers of impoverished Indian women, 'a woman has to have 6.2 children to have at least *one surviving male*' (Desai and Krishnaraj, 1987: 243). While such conditions and high infant mortality are most characteristic of the north Indian states of Uttar Pradesh and Rajasthan, they are found throughout the subcontinent, wherever there is extreme poverty. These material conditions of existence—rather than simply the desire to have a male child—are at the root of the population problem. Mamdani was among the first to note the paradox that among the poor, considerations of family size, power, and chances of employment for males provide great incentive to produce more sons (1972; see also Goldscheider, 1971). Recognizing that life is 'haphazard', i.e. out of one's control, the poor must invest in several sons to ensure that at least one may obtain an income. And of course, it is among the poor that the material conditions of existence are most precarious and so these women suffer most from the impact of multiple pregnancies. Further, as Mukhopadhyay notes, women of childbearing age are 'dying in greater numbers *despite advances in health services and medicine*' which might be explained by the increasing inequitable distribution of such services (1984: 28; my italics). The difficult conditions under which women live and work in the struggle for existence, combined with a lack of basic knowledge regarding health, result in exhaustion and helplessness—especially for poor women.

All but four of the twenty-one women who spoke specifically about their reproductive histories had lost children (although all miscarriages were not recalled), and these four came from relatively better-off families of Panjabi Jats and Ahirs. A pattern emerges from their collective experiences, however, in which not only did they continue their daily work routine right up until the time of delivery, but their first childbirth was marked by a critical lack of knowledge about what they were to undergo. In the case of one woman, this had the most tragic—almost unbelievable—consequences.[24] But for the others as well, this first delivery occurred neither with adequate knowledge nor under adequate conditions of care and sanitation.

I didn't know anything about childbirth when my first child was born. I called my nanad (husband's sister) the first time to help. When she came, she asked me to work at grinding the flour at the grindstone (*chakki*). My pains had started and I couldn't sit there. But I didn't realize they were labour pains, so I only told her, 'No, I can't sit at the chakki.' She became really angry and pushed me away, saying, 'You go back to your parents' house (pihar).' I just swallowed my anger and went to the jungle. I came back with fodder for the animals. I didn't eat or drink anything and my nanad didn't give me any roti. I fed the animals; then the pains really started. I wanted to go into the house but my nanad wouldn't give me the key. I fought with her (and finally got the key). I went into the house and my niece, who was returning from grazing the animals, saw me and then went to get some water and lay down on the cot. Then I told her to go and fetch somebody. Some women came to help me. I sent for my nanad; she came and started cursing me. The other women asked her to keep quiet and then I gave birth to a girl. Then they called the midwife (*dai*) who cut the cord. My nanad can't adjust with anybody.

I didn't know anything about how a child is born. Well, everybody knows this much—that a child comes from the front. That is a feat of God. It is produced from a man and a woman being together.

The whole day I worked in the fields and in the night I had a baby. The first time, I ground 10 to 11 kilos of flour on the chakki. While doing this I felt pains, but I continued my grinding anyway. As soon as I got up, I could no longer bear the pain so I woke up my sas and told her, 'I have some pain. Please come with me to the field (to the latrine).' My sas said, 'No, you're not telling the truth. You don't have pain for going to the latrine, but for something else.' She sent my younger brother-in-law (*devar*) to call the dai. I told her I had a terrible pain in my stomach. My sas told me to push hard. I couldn't do much and sometimes I stood up when the pain was unbearable. She gave me *ajwain* (anise) and *gur* (hard molasses) with water and she let me sit down. Soon after this, my daughter was born.

I didn't know anything about childbirth before my first child. I had a child about one year after marriage. I was really afraid. All my relatives were around when my children were born; but I didn't call any of them. I didn't want anybody to know because I was too shy. My bhabhi (who was a trained midwife) was there, but I didn't call her. I didn't even call the dai from our village. All three children were born only with the help of my sas. I never used to sit in front of anyone when I was pregnant; I used to stay inside. I didn't even sit with my bhabhi and brother.

When I was first pregnant, I returned to my pihar. I wouldn't eat anything. My bhabhi asked me what's the matter. I said I didn't feel anything (wrong). She asked me when I had last had my mahina. I told her three months back; she told me I was going to have a child and not to tell anybody but my sas. Then I went to a fair (*mela*) and didn't feel well there. I started crying and my husband asked me why and I told him. That's how my sas got to know. In the beginning I used to feel shy. They called the dai (for the delivery). When my twins were born I went to the hospital in Alwar. I had had one premature baby and when the doctor came he said I must go to the hospital. The twins were born when they were six and a half months. I was going to the fields (at the time) to bring fodder when I slipped on a stone and got hurt. Then the pains started and when I returned home they called the doctor.

I knew nothing before the first child. No one was called that time and my sas took care of everything. The child was born in the fields. I didn't know anything about it (childbirth). I was eating some roti there and an old woman said, 'Show me your stomach', and then she told me to go home. I returned with a heavy load on my head. Then I wanted to go to the latrine and again went to the fields. I told my nanad to call mother—I have pains. My sas reached and saw the child's head coming out. She told me to come home and that I could go to the latrine there. Then she brought a *parat* (shallow, metallic vessel for kneading flour), and the child was born. My fourth child was born in my pihar. My brother fetched the dai there. I was lying on the cot when the child was born. I had terrible pains and by the time the dai came the child was born.

After delivery, a mother requires adequate rest and nutritional intake not only to regain her own strength, but to pass on needed nutrients to the infant. The composition and income of the family are crucial in determining how much rest and food a woman can receive. When households are joint, and there are other women in the family (mother-in-law, other bahus, or husband's sisters), the chances are greater that a woman will be somewhat relived of her daily chores for the needed rest.[25] When the family is nuclear, as was usually the case with the women interviewed, rest became a matter of the availability of some relative to come and help with the housework and other childcare. Noteworthy are two instances where women said that their husbands cooked the food and did all the work (one was Chamar and one was a Bhangi). Some women started working five days after their deliveries—others after ten to fifteen. None *felt* they had sufficient time to recoup their strength. The special foods that are desirable for the postpartum mother, especially after her first delivery (such as noted by Jeffery *et al.* 1989: 129), remain a luxury for poor women. Even ghee is often not available and we were told that some were given a porridge of crushed wheat with molasses (*dalia* with gur) or sweet porridge of wheat flour (*atta ka dalia*). The women felt they had inadequate rest and inadequate food.

All the women expressed great joy when their first child was born—whether male or female; 'when a child is born—at that time everyone thinks only about its life'. There is an especially close relationship between the mother and daughter, for two main reasons. It is the young daughter who, at about five or six years of age, begins to learn to perform the myriad household chores. In poor families where the female children are not sent to school, they assume a large measure of these tasks by the age of eight or nine.[26] Time and again village women said that it was only their daughters who (even after marriage) really cared for them and would help them—not their sons. No woman thinks her household is complete without a daughter.[27] Although the Hindu custom is to hold a function upon the birth of a boy (*kua puja*—worship at the well and provision of a feast to relatives and villagers), this was often not done because of poverty. Panjabi Jats held a *hawan* (sacred fire worship) or would feed some of the *kanya*s (pre-puberty girls) in the village. On the other hand, a special dish of rice and other food might be served to relatives to mark a girl's birth.

V. FAMILY PLANNING: KNOWLEDGE AND REALITY

Shankpur village has an extremely high number of inhabitants who have undergone sterilization as a family planning measure. In practically all these cases it was the woman who had a tubal ligation. None of the women—except for the generally more educated and 'progressive' Panjabis—knew of any of the government-advocated measures that would enable them to space their births; neither pills, the loop, nor condoms. The only thing they had heard about was the operation.[28] Although we did not pursue the matter, it is possible that some women know of contraception and also reject it, for it removes the only legitimate reason they have for controlling their sexuality—i.e. fear of pregnancy is a major weapon for not having sex and, alternatively, for having it (see further, Morokvasic 198; compare Nabeer, 1985: 85–6).

The only knowledge available to women for spacing children appears to be that of abortions, which has also been a part of traditional forms of fertility control. Pills are also available from the village dispensers of medicine (locally-trained 'doctors') that may have the effect of inducing menstruation; these are often sold to women as abortifacients, with the advice that it will stop their pregnancies. There are two options open to women—and just one for poor women—other than having an unwanted child. The first is to have an abortion in the government hospital, which may, at a premium, cost as much as Rs 400. The second is to take some local (*deshi*) mixture which will induce abortion by its hot (garam) qualities. Generally, it appears that Ahir women are the least likely to have abortions. And, although it is not as common among Meos—as among the Untouchables—poor Meo women may resort to this for limiting pregnancies. Panjabi women generally have an operation as soon as the 'desired' number of children is reached.

Previously women could go to *vaidya*s (*munshi vaid*, homeopathic doctors) if they wanted an abortion and would take some *deshi kardhva* (bitter, local/homeopathic) medicine. This was a combination of all hot things—*gajar ka bij* (carrot seeds), *tulsi ka patta* (*tulsi* tree leaves), old *gur* (molasses), *til* (sesame), and *ajwain* (anise). It was all mixed and you drank it. If a woman didn't know of this she would learn about it by talking to others. Some woman may still take this. How can poor people like us afford this much money (for government abortions)? (Then an example was given of how a woman saved some 600 rupees by being provided with this information for her daughter-in-law's abortion). After two days her *mahina* started and a big clot of blood came out. It cost only five rupees to purchase this stuff.

After the birth of my second girl I was again pregnant and had an abortion in the third month. I told my *jethani* I couldn't manage with so many small children. I asked my husband (for permission to have an abortion), and he didn't tell my *sas*. Later (when she found out), my *sas* said, 'Why did you have abortion? Perhaps you could have had a son.' But I thought it better. Because of this longing for a boy we had so many girls (three at the time of the interview, and a fourth later). The abortion was done in the Alwar government hospital. It cost about Rs 400 which my *jethani* gave. I don't know how or what was done; they gave me anaesthesia and I became unconscious. After three months there is life in a child. That's why they don't do an abortion after this period. If women can't go to the hospital for an abortion, they have the child. That's why old people have so many children. Even now my *sas* talks about it and says it could have been a boy.

I went to get an injection from the village compounder one and a half months ago. My *mahina* had stopped. I was taking the pill but it didn't work out. He guaranteed that my *mahina* would surely start after having the injection, but it didn't. I gave him ten rupees and the second time I gave him six. Then he said, 'I can take you for an abortion and then you can also have an operation.' But then I will have nothing if I do both things at once (*mere pas to kucch bhi nahin rahega*). First I will have the abortion and then the operation (e.g. I will have neither my child, nor will I be able to bear another). I will die from my weakness (*main to mar jaungi*). My husband had an operation and he is finished (weak and unable to produce children).[29] I got another injection to start my *mahina* and it came only half a month later.[30]

Despite the fact that sterilization is less risky when performed on men, it is almost universally the women who undergo this operation. Women told of varied experiences—mostly about being forced into doing it when their husbands refused to have vasectomy, and some of having to do this surreptitiously on their own. In the latter situation, the women exhibit considerable courage and exercise personal autonomy. But these cases are rare and do not represent a common form of resistance. Still others told of hoping to have an operation after one more son. Village women also have internalized the male evaluation of their own work output and contributions to household survival. They belittle how much and how hard they actually work.

My husband decided about the operation—that I have it. He said it will be better if we have a small family and we can educate them and bring them up properly. *He* would not get it as the operation is sometimes unsuccessful and the wife gets pregnant. Then people will ask whose child it is. Also, it can ruin the man's health. He has to go out and work, go to the mountain and cut stones (do *mazdoori*). He thought he may become weak and so asked me to do it. (And if you get weak?) I only earn little bit; he is the major earner. He did say he would get the operation, but I told him not to. I stopped him as he does all kinds of work—ploughing, watering, digging. I can only do a little bit of work in the fields and look after children.

I went to Alwar for an operation. A nurse from Rampur (a nearby village) used to come here. When I was pregnant with the last girl, I told the nurse that I will get an operation after this delivery. She kept coming to see me often and to make sure her (prospective sterilization) case was *pakka* (real).[31] She found out when the child was born and came within four days. I never told my sas—what's there to ask her about this! I was telling my husband to go and have the operation but he said, you go. We were both afraid. Then we started fighting when the girl was two to three days old. I told him—if I have to die because of the operation, then I will leave the house—but I won't have an operation. He told me not to think this way; others have had an operation and didn't die—they won't kill you. I asked why he wouldn't go. He said, suppose he gets sick, then a woman can't do mazdoori (ironically, she later became a widow). That's why he told me to get it and that even if I get sick he'll take care of me. He said I could rest at home. We considered all these things and finally I went. I told him, 'If I die, you take care of the kids.' After four days (of the birth) I went with the Rampur nurse; I was so mad at him, but I couldn't help it. I was unconscious for the whole first day; the second day I became conscious. The nurse had taken me to Alwar and I had the operation there at 9 in the morning.

I had an operation after only two kids. My husband told me to get it done. I didn't really want to, as I wanted a daughter. But my husband insisted on it. He said that we don't have enough land to support a large family. I used to have difficult deliveries, so we decided on the operation.

My in-laws and husband were against it, but I insisted. I told the village (trained) midwife that I wanted an operation and she took me. I came back after 8 days. My husband was very angry. He said, 'Who told you to go for the operation. Now go from my house—I will not keep you.' I said, 'No, you can't take care of the kids. So you go if you want but I won't leave this house.' Then I explained to him that we already have four children and should not have any more. If they are healthy then it's fine (to have the operation). And if you want more—bring another wife. We cannot even support these four. My *sasur* (father-in-law) was angry but my sas said nothing.

I am expecting now and will have an operation after this birth. My husband wants me to get it. He drinks a lot and the doctors say that you have to give up alcohol before having a vasectomy. (She told the story of a man who drank before his operation and then died.)

If my fourth child is a daughter I would like to have an operation. Yes, in fact, I would like one now. Our fate is such that we have so many girls (she has three). But my nanad, sas, and sasur will not let me have it done. They want me to try once again (for a boy), and then have the operation done no matter whether it's a boy or a girl. If I had just one son I would never have had more than two children.

I haven't had an operation yet. I'm trying for another boy and then I will get it. We'll wait and see. My husband is not very strong so probably I'll get it. (But you are not strong either. Why do it?) Well, people put a lot of pressure on us and we have to do it. What can we do? My husband does not want it done. He doesn't say he will go and doesn't say I should go. But the government is forcing people to get an operation done. (But it may affect your health—you are weak.) I can manage in the house. I don't work much. I only recently started working as much as my husband. If men have to work hard, they should be strong. If I tell my husband to go for the operation he will agree, but I don't want him to.

My husband refuses to do it. He says, if you want to—you do it. I will have it done after my third child.

The most striking thing about the above quotations is what they reveal of the contradictions inherent in women's position in the household. On the one hand, they are desirous of birth control, and on the other they are the repositories for reproduction of the family unit and the requisite number of sons. On the one hand, they ideologically put little value on their own labour; on the other hand, both women and men admit the former work harder and are critical to household survival. The few cases where women acted with extraordinary courage to defy the wishes of others are, unfortunately, rare occurrences.

CONCLUSIONS

An analysis of reproductive activities in terms of the social relations of reproduction emphasizes the dynamics of social divisions. At the most basic level lies the gender division which infuses the sexual division of labour in society. The costs and benefits of children, therefore, are not shared equitably among different members of the family. Household members also have differing claims over its resources and such claims 'reflect culturally-determined notions about needs and contributions' (Nabeer, 1985: 84–5; see also Batliwala, 1985). Nabeer found in Bangladesh that

adult males have a prior claim on consumption resources because they are regarded as breadwinners, guardians, and protectors. Women's claims vary with the position they occupy in a particular household at different stages of their life cycle, while children's claims are likely to vary with age and sex (1985: 85).

In class-based societies, the social relations of gender are entangled as well with the divisions based on class. Women are affected in a specific way by fertility decisions as they carry the heaviest burdens of childbearing and childrearing. As Beneria and Sen note,

> multiple pregnancies affect the mother's health, work, well-being and ability to participate in activities outside of the house. Poor peasant households may survive from the continuous pregnancies and ill-health of the mother, exacerbated by high rates of infant and child mortality. Here, the contradiction between class interests and their role as women becomes evident (1986: 155).

Women are also left with the everyday task of feeding and providing for all the children as their husbands are often away, either working or looking for work. For the poor women, there is no assurance that the husband will provide (send) them the money needed to feed the family.

The social and political relations of reproduction within a Rajasthani village household, in which women have little prior knowledge and often less control over their own reproductive activities, reflect the subordinate position of women. This is especially true for the young girl and her continuing socialization as a new bride as she encounters menstruation, sexuality, childbirth, and fertility. It highlights a pervasive lack of control and autonomy over women's own generative capacities and sexual life which, while to a large extent cutting across classes, has different consequences depending upon the individuals concerned as well as the specificities of their own existence. In drawing attention to the dual role of reproduction for the continuation of a social formation—as both reproduction of labour power *and* reproduction of the social order itself—Krishnaraj foregrounds the role of socialization and subordination of women in that process (1989: 24). As replication of a social formation depends upon the socialization of its individuals, the primary purview of women, the latter are expected to act in all ways to ensure this reproduction—

not the restructuring—of the social formation. Women are also charged with upholding the family's status through their actions and this further embeds an inherently conservative bias in defining the parameters of what women may or may not (should or should not) do to ensure continuity of the social order. The social relations of reproduction also indicate how women's role in production is relegated to secondary status through the sexual division of labour in the family.

The methods, goals, and control of women's reproductive capacities is not only clearly a contested terrain—it is a contest among unequals. Of particular importance is the documented manner in which women's subordination is 'reproduced' in the wider arena of social relations, in this particular context in the arena of state policy on population control. These policies focus on sterilization of women as the almost exclusive means of controlling fertility rather than attempting any transformation of the material conditions affecting women's existence and the need for children, as the fertility decline in Kerala reveals (e.g. Ratcliffe, 1978; Mahadevan and Sumangala, 1987; Franke and Chasin on Kerala for an example of a different route to fertility control).

Using the example of the politics of family planning in Brazil, recent research provides an opening to reconsider the questions of fertility control and women's autonomy.

In several articles dealing with differential fertility and distribution of per capita income in north-east Brazil, Daly found that one important reason for the income disparity between the poorer 80 per cent of the population and the richer 20 per cent, and 'the absence of any net "trickle down effect", was the more rapid population growth of the lower class, due to much higher fertility which more than compensated for the higher mortality of the poor' (1985: 329). He suggests adding to the Marxian notion of exploitation based on class monopoly of the means of production—

a notion of Malthusian or 'Roman' exploitation based on *class monopoly of the means of limiting reproduction*. In ancient Rome (as in north-east

Brazil) the role of the proletariat was to procreate a plentiful supply of labourers and servants for the republic (i.e. the patricians) (1985: 329, my emphasis).

Daly asks us to suspend the Marx-Malthus dichotomy in favour of a definition of social class in terms of control/non-control of both production and reproduction. He traces a significant decline in fertility among the poorest, despite their continuing immizeration, thus discounting the usual 'demographic transition thesis'. Instead, he sees this as a result of a certain democratization of birth control in terms both of attitudes and access to devices. He concludes that 'the current democratization of control over reproduction is no less (and no more) important than the democratization of land ownership in the north-east. Everyone talks about land reform, but so far few talk about "reproduction reform"' (1985: 338)—and paradoxically, it is the latter which seems to be occurring more rapidly.

A concrete example of the struggle of poor women in Brazil to control their own sexuality and to demand access to birth control comes from Barroso and Brushini's study in San Paulo and the action research project on women's sexuality conducted by the Chagas Foundation (1990; see also Accad, 1990).[32] The project originated in demands for discussion on sex education and gender relations voiced by women in grassroots movements, as well as the contradictions surrounding the history and politics of pronatalist population policies in Brazil. 'The aims were to construct a collective knowledge about sexuality and to share this knowledge immediately among all the participants'— to discuss the meaning of sexuality in their intimate relationships as well as the broader social context within which they occur. One objective was particularly geared towards enabling mothers 'to accept and respond positively to the sexuality of their children, in order to *avoid the reproduction of ignorance and shame of which they themselves were victims*' (1990: 161, my emphasis).

Barroso and Brushini also point out the dilemmas of birth control for the left, in which their position has become the mirror-image of Malthusian thinking. The left rejected birth control per se, rather than seeing it as an adequate response to particular historical circumstances.

The class interests of the proletariat were sometimes defined merely in terms of the negation of the interests of the bourgeoisie. The economic advantages of large families observed in some contexts were uncritically generalized as if they applied to all sectors of the poor, rural and urban alike, and the costs of high fertility to women and children were ignored, as if they were shared equally among all members of the family (1990: 157).

At the same time, working with the women permitted researchers to rethink the relations between the individual and the broader structure of social relations. Although remaining clear in their understanding that birth control 'could not be used as a solution to economic problems ... [and] the primary cause of poverty is structural', they were gradually convinced that 'information about and access to contraception are also important elements for the improvement of their living conditions on a short- term basis' (1990: 167).

The contradictions inherent in the problematic of unpacking gender and reproduction that have been analysed here reveal themselves in a multi-layered complexity. On the one hand, they are linked to questions of the reproduction of the social formation and the extent to which women may or may not be able to resist their subordination. In looking at the comparison between the situation of a rural area in India, such as Shankpur, and poor women in San Paulo, it appears that the question is not so much why women don't resist, or even the 'everyday forms of resistance' that appear to be favoured 'weapons of the weak'. Instead, we need to learn more about the still rare, but increasing, conditions under which women do come together to take concerted action. The Chagas Project has much to tell on this account, when individual forms of resistance may come together to make a directed change.

On the other hand, the contradictions are also part of the large issue of the capitalist development strategy to which the Government of India has committed itself. From the pioneering work of Boserup (1970) until today, study after study has revealed how women's integration into the 'development' process writ large has served to deepen their subordination. But here, it must be carefully noted that important differences exist between women in different classes. Poor women have been 'integrated' in the least advantageous position. Development strategies, built upon a patriarchal society, have adversely affected their workforce participation and role in production processes, their nutritional intake (Mukhopadhyay, 1984; Sharma, 1989), and their control over their reproductive capacities. Yet it is not clear how women—especially poor women —will continue to deal with the competing needs of reproducing to ensure household survival and of controlling fertility in order to ensure their own survival. One thing, however, is certain. As the conditions of the poor become more and more marginalized, impoverished, and insecure, the contradictions between the needs of women and the household will become even sharper.

NOTES

1. Mackintosh develops the concepts of subordination and the household in depth (1981: 1-15). She points out that in developed capitalism the household mediates two sets of social relations—those of marriage and filiation and the wider economic relations of society. The former both constitute the household and determine the context of much of child care. Women's domestic work, she finds, is the economic expression of the fundamental inequality within marriage. She concludes that the subordination of women through the unequal division of labour in the wage sphere is ultimately *derivative* from their subordination within the marriage-based household (11-12).
2. Mukhopadhyay attributes this to the historical destruction of household industry brought about as a result of colonization and structural changes in the economy (1984: 30).

3. There is a growing debate over the relationship of poverty and population, framed within the issue of the value of children (economic and otherwise) as a determining factor. First raised by Mamdani's 1972 village study (restudied by Nag and Kak, 1984 to note the economic changes that have reduced the economic value of children, while increasing contraceptive use), it has been taken up by Cain in an extended series of articles (1977, 1980, 1981, 1982, 1983, 1986). The counterattack has been led by Caldwell (especially 1976 and 1978; also 1980, 1981, 1983) and others (e.g., Cleland and Wilson, 1987) whose Chayanovian vision of peasants has led to a focus on fertility decline due to social factors affecting *intrafamilial* wealth flows, the transformation of values, and spread of education. Robinson takes on Cain's high fertility as risk-insurance thesis to conclude that 'it simply does not hold up, either logically or empirically'. He concludes that the winner is not yet declared in the 'development against population control debate' (1986a: 298, 290). In fact, his affirmation of Day's '"drift" behaviour model to cover such very poor, apparently planless life-styles' supports a strong population control policy. Cain's rebuttal (1986) is instructive; neither, however, contextualize demographic changes within the context of the contradictions of capitalist development in Third World societies which —amidst an increasing proletarianization and pauperization—also produces population growth (see, for example, Tilly 1984). Another series of articles by the Vlassoffs (1979, 1980, 1982) takes an anti-Mamdani *et al.* position. They posit that people do not have children either because they are an economic asset (as there are no jobs) or provide old-age security. The critique by Datta and Nugent (1984) brought about a rejoinder (1984) and subsequent negation of the value of sons even for widows (Vlassoff 1990).

Finally, there is the debate regarding landholding and fertility, where children may be regarded as a security asset that complements, rather than substitutes for, land (Cain 1985, 1986; Stokes *et al.* 1986). Cain also calls for a broadening of the inquiry regarding the determinants of fertility beyond empirical or 'narrowly focused theories of household behaviour' 'to sketch in the institutional reality of particular settings, and to incorporate this reality in analyses of individual reproductive behaviour' (1985: 15). On a macro-level, on the other hand, Dyson and Moore posit a model of north-south gender relations which echoes Mitra's (1978) suggestion that female social status is probably the single most important element in comprehending India's demographic situation' (1983: 54).

4. The unequal sex ratio is discernible from the early decades of this century and has *accelerated* since independence. 'Women's lives are cheaper and more expendable than men's' (Mukhopadhyay, 1984: 25; see tables on p. 26; also Desai and Krishnaraj, 1987: 223ff).
5. Such information is deeply personal and private; the responses varied for each women, as did the extent to which we felt comfortable in pursuing the issue, and are not to be considered 'complete'. We also felt constrained from following up with questions at certain points. The extent to which it may reveal a reality beyond the individual women concerned can only be attested by further research.
6. Substantial parts of this section are taken from Sharma and Vanjani, 1989. The name of the village is changed to respect the privacy of those who shared the intimacy of their lives with the authors. We lived in Shankpur from October 1986 to May 1987. The research was supported by a Smithsonian Senior Foreign-Currency Award to M. Sharma as well as by numerous friends in India. Special thanks go to Lori Yamamoto for her help and patience in transcribing notes.
7. While the term 'Jat' refers to a caste, it is also freely used in Panjab to include all those whose profession is agriculturist.
8. Alwar is in the area traditionally known as 'Mewat'—land of the Meos. Before the holocaust of the Partition of the subcontinent, they were the largest and most dominant group in the village and the region. The majority left for Pakistan.
9. There were even two cases when a woman and a young girl had to plough because no male was available in the household at the time (in the former case, the husband was in jail; in the latter the father had died).
10. There are, of course, extraordinary times when a man may assume household and childcare duties. These occur if his wife has given birth or is seriously ill and no other female is available.
11. Other studies also list physical tiredness, headache, and faintness as the most common ailments women describe, as well as chronic anemia due to blood loss, and protein and calorific deficiency (described in Chanana, 1989: 150, 154).
12. These were trained as medical aides by the Rajasthan government.
13. Sharma also soon became known as a dispenser of medicine in the village. Those who came to her were the poorest, mostly women. With a small arsenal of aspirin, antibiotics, various skin ointments,

dressings, and dysentery pills (refusing to treat anything serious, as this was beyond her capabilities) she was in a position to learn much about these problems. She also learned much about what she could not treat.
14. Drinking was far and away the number one social problem afflicting the village. It affected all castes and all classes.
15. One woman said that her older sister explained what happened when she asked her, 'where did the blood come from?'
16. The translation of *sharam* as 'shame' would be derived from its context.
17. Aziz found that girls as young as 8 or 10 years old in rural Bangladesh begin to wear a 'scarf' (*ornal*) to cover their breasts from male eyes (1989: 59).
18. While the age at marriage for non-Panjabis was low—Rajasthan is known for the highest incidence of child marriage and five or six years of age is not a rarity—that of the Panjabis was considerably higher, generally between 15 and 17 for girls.
19. Aziz also found that 43 per cent children between 6 to 9 years old slept with their parents, but that no child over 9 shared the bed (1989: 66). Kapur also found in his unusual study of the sexual life of rural Kumaonis in the northern hills of India that young children see their parents copulating (1987: 29).
20. Not during these interviews, but on other occasions I have heard (from women and men) that the first night was more like a rape. Kapur also relates moving stories of wedding nights from male and some female villagers. He found that in all castes, the men and women did not perceive their wedding night as memorable. Most informants felt it had been troublesome, even horrible. To some of the women it had been a night when the brute strength of husbands had triumphed. One experienced a severe beating. He concludes, 'the wedding night then is usually a traumatic entry into sexuality for most Kumaoni women. For the men, it is an unsatisfactory introduction to their wives' (1987: 54; 42ff).
21. One woman is reported by Kapur to have said, 'when I married I was an ignorant girl. I did not even know how children were born. I used to imagine that when a man touched you, you would begin to have children ... my husband was not experienced and our first night was one of horror for me' (1987: 42).
22. One woman told Kapur how she tried to fight off her husband on

her wedding night, and even shouted to her relatives to help. 'But it is the custom to ignore such shouts on the wedding night' (1987: 47).
23. One Kumaoni woman said that 'my girlfriends had told me that it would be great fun but I was only hurt and had no enjoyment' (Kapur, 1987: 42).
24. Her story is included in the notes because it is so atypical. She narrated that when she became pregnant the first time, she knew nothing about how childbirth occurs. Her husband was at work and she became nauseous. She went to the compounder ('doctor') who gave her two pills; one for the morning and the other for evening time. She swallowed one pill. In the evening, while cooking, she started to bleed but she didn't feel anything. She was wearing a white petticoat. When her husband came home, he saw the petticoat all red and asked, 'what have you done?' She stood up at that time and felt some pain. She went outside her house and squatted by a drainage ditch where two children were born—both alive. But, she continued,

we didn't know anything, so we took them and put them in a box and buried them in the ground. ('Couldn't you see that the children were alive?') We didn't know that (they were alive); we thought that since they had fallen down (into the ditch, *nali*) there was no more life (*janam*). Then I became unconscious and didn't know what happened.

My husband went to the doctor and told him what had happened. He came to see the children; as we had closed the lid, their breath had stopped. The doctor said that they were dead and so they buried them (again). Then my doctor took me to the hospital. I was admitted, still unconscious, and only became conscious on the next day. As I had never seen a hospital before, I started crying, thinking my husband sold me (to some place). The doctor came and told me not to cry—my husband would be coming soon. When he came I asked him why he had left me there all alone.

Then they took off all my jewelry and my husband said that they were taking me inside for an operation. I insisted on going back home, but he said I couldn't until the doctor released me, otherwise they might put me in jail. ('What kind of operation was that'?) You see, the doctors didn't know the children were born at home. He pressed my stomach and gave me some bitter medicine to drink. I was there for two months. And the pill the compounder had given was the wrong one—it was *garam*. He was suspended from his service for this.

25. Only one woman told of being able to get one and a half months' rest after the joint-family was divided. 'When we lived together there was no rest. The fifth day after delivery I ground wheat and *bajra* and did all the other work. My sas had separated us after my first child was born. Then I called my *bua-sas* (mother-in-law's sister) and she helped do the work. The next time I called my sister and I got a rest for 3 months. The next time I called my sister again and could rest for two to three months. With the last child, my husband did all the work and the children did little jobs. My husband helps a lot (even now).
26. Boys were much less likely to be withdrawn from school or prevented from attending in order to help perform agricultural tasks or provide a supplement to the family income. Only the poorest households kept their sons out. Universally, a boy's education is regarded as an important investment whose dividends are reaped in the future.
27. An unique case was that of an Ahir woman who said that, after the birth of two sons, her husband forced her to get an operation. He said they don't have enough land to support a large family.
28. It is not clear whether this is the case for men as well; that is doubtful. However, several women did say that neither they nor their husbands knew of these matters; they were 'simple'. On the other hand, as Morokvasic notes for Yugoslav migrants as well, there is a reluctance to use contraception on the part of *both* men *and* women. This stems, she writes, 'from the way in which relationships between a man and woman are structured, and in particular women's position with this relationship.' Some men mentioned open resistance to using contraceptives as related to less sexual desirability. Many of the women had internalized men's reasons for resistance: they would become 'free' too; become 'like a man' and sleep with anybody; 'no longer be faithful'. The women resented one of the most important possible effects of contraception—'the disruption of a deeply embedded authority structure and the double standard in which the man is dominant and sexually free while the woman is subjected and faithful' (1981: 135).
29. Her pregnancy was by another man.
30. The woman whose experience is described here is an extraordinary one. Her husband is a much older, ailing man who often stayed in Alwar to work. She had taken a lover from another *dhani* in the

village, with whom two children were produced; a year later she was pregnant with a third. On one occasion, some male elders in her caste brought the issue up at a village council (*panchayat*) meeting and stated that the situation was intolerable; however, they have been unable to stop it.
31. There is considerable pressure for government workers to meet quotas of sterilization cases.
32. The project was funded by the Ford Foundation, 'which has supported many progressive initiatives in Brazil, including some of the most radical feminist groups' (1970: 170). This was also possible 'because the women of the poor peripheries of the cities asserted their needs, with no respect for the hierarchy of needs established by the well-meaning but misguided "bread first" school of thought' (1970: 171). An earlier, thoughtful piece by Bondestam (1980), while basically presenting a Marxist analysis that links fertility control to socio-economic conditions, also recommends not population control—where people are controlled by others—but family planning that people control themselves. A woman should be given the opportunity to limit her number of children just as 'she should also be given the right to education, a productive job, stable economy, a meaningful social standard and other fruits of well-planned development' (1980: 36).

BIBLIOGRAPHY

Evelyne Accad, 'Sexuality and Sexual Politics', in Chandra Mohanty, Ann Russo, and Lourdes Torres, ed., *Third World Women and the Politics of Feminism*, Bloomington and Indianapolis: Indiana University Press, 1991, pp. 237–50.

K. M. A. Azziz, 'Daughters and Sons in Rural Bangladesh: Gender Creation from Birth to Adolescence', in Maithreyi Krishnaraj and Karuna Chanana, ed., *Gender and the Household Domain, Social and Cultural Dimensions*, New Delhi/Newbury Park/London: Sage Publications, 1989, pp. 55–73.

Srilatha Batliwala, 'Women in Poverty: The Energy, Health and Nutrition Syndrome', in D. Jain and N. Banerjee, 1985, pp. 38–49.

Carmen Barroso and Cristina Bruschini, 'Building Politics from Personal Lives, Discussions on Sexuality among Poor Women in Brazil', in Chandra Mohanty *et al.*, 1991, pp. 153–72.

Lourdes Beneria and Gita Sen, 'Accumulation, Reproduction, and Women's Role in Economic Development: Boserup Revisited', in Eleanor Leacock and Helen I. Safa (contributors), *Women's Work: Development and the Division of Labour by Gender*, Amherst: Bergin & Garvey Publishers, Inc., 1986, pp. 141–57.

Lars Bondestam, 'The Political Ideology of Population Control', in Lars Bondestam and Staffan Bergstrom, *Poverty and Population Control*, London, New York, Toronto, Sydney, San Francisco: Academic Press, 1980, pp. 1–38.

Ester Boserup, *Women's Role in Economic Development*, London: Allen & Unwin, 1970.

Mead Cain, 'The Economic Activities of Children in a Village in Bangladesh', *Population and Development Review* 3(3): 201–27, 1977.

———, 'Risk, Fertility and Family Planning in a Bangladesh Village', *Studies in Family Planning*, 11: 219–23, 1980.

———, 'Risk, and Insurance: Perspectives on Fertility and Agrarian Change in Rural India and Bangladesh', *Population and Development Review* 7(3): 435–74, 1981.

———, 'Perspectives on Family and Fertility in Developing Countries', *Population Studies* 36(2): 159–172, 1982.

———, 'Fertility as an Adjustment to Risk', *Population and Development Review* 9(4): 688–702, 1983.

———, 'On the Relationship between Landholding and Fertility', *Population Studies* 39(1): 5–15, 1985.

———, 'Landholding and Fertility: A Rejoinder', *Population Studies* 40(2): 313–17, 1986.

J. C. Caldwell, 'Toward a Restatement of Demographic Transition Theory', *Population and Development Review* 2(3–4): 321–66, 1976.

———, 'A Theory of Fertility: From High Plateau to Destabilization', *Population and Development Review* 4(4): 553–77, 1978.

———, 'Mass Education as a Determinant of the Timing of Fertility Decline', *Population and Development Review* 6(2): 225–55, 1980.

———, 'The Mechanisms of Demographic Change in Historical Perspective', *Population Studies* 35(1): 1–27, 1981.

———, 'The Causes of Marriage Change in South India', *Population Studies* 37(3): 343–61, 1983.

Karuna Chanana, 'Introduction, Gender Dimension of Health', in M. Krishnaraj and Karuna Chanana, 1989, pp. 145–55.

John Cleland and Christopher Wilson, 'Demand Theories of the Fertility Transition: An Iconoclastic View', *Population Studies* 41(1): 5–30, 1987.

H. E. Daly, 'Marx and Malthus in North-East Brazil: A Note on the World's Largest Class Difference in Fertility and its Recent Trends', in *Population Studies* 39(2): 329–38, 1985.

S. K. Datta and J. B. Nugent, 'Are Old-Age Security and the Utility of Children in Rural India Really Unimportant?' *Population Studies* 38(3): 507–9, 1984.

Neera Desai and Maithreyi Krishnaraj, *Women and Society in India*, Delhi: Ajanta Publications, 1987.

Tim Dyson and Mick Moore, 'On Kinship Structure, Female Autonomy, and Demographic Behavior in India', *Population and Development Review* 9(1): 35–60.

T. Scarlett Epstein and Darrell Jackson, *The Paradox of Poverty: Socioeconomic Aspects of Population Growth*, Delhi: Macmillan Co. of India Ltd, 1975.

W. Richard Franke and Barbara Chasin, *Kerala: Radical Reform as Development in an Indian State*, Food First Development Report No. 6, San Francisco: Institute of Food and Development Policy, 1989.

Calvin Goldscheider, *Population, Modernization and Social Structure*, Boston: Little Brown & Co, 1971.

Linda Gordon, *Woman's Body, Woman's Right: A Social History of Birth Control in America*, New York: Penguin Books, 1977.

Devaki Jain and Nirmala Banerjee, *The Tyranny of the Household: Investigative Essays on Women's Work*, Delhi: Shakti Books, 1985.

Patricia Jeffery, Roger Jeffery and Andrew Lyon, *Labour Pains and Labour Power: Women and Childbearing in India*, London/New Jersey: Zed Books Ltd; New Delhi: Manohar, 1989.

Naila Kabeer, 'Do Women Gain from High Fertility?' in Haleh Afshar ed., *Women, Work, and Ideology in the Third World*, London/New York: Tavistock Publications, 1985, pp. 83–106.

Tribhuvwan Kapur, *Sexual Life of the Kumaonis, A New Approach to Sexuality*, New Delhi: Vikas Publishing House, 1987.

Maithreyi Krishnaraj and Karuna Chanana, ed., *Gender and the Household Domain, Social and Cultural Dimensions*, New Delhi/Newbury Park/London: Sage Publications, 1989.

Maureen Mackintosh, 'The Sexual Division of Labour and the Subordination of Women', in K. Young, *et al.*, 1981, pp. 1–15.

K. Mahadevan and M. Sumangala, *Social Development, Cultural Change and Fertility Decline: A Study of Fertility Change in Kerala*, New Delhi/London: Sage Publications, 1987.

Mahmood Mamdani, *The Myth of Population Control*, New York: Monthly Review Press, 1972.

Mirjana Morokvasic, 'Sexuality and Control of Procreation', in K. Young, *et al.*, 1981, pp. 127–43.

Moni Nag and Neeraj Kak, 'Demographic Transition in a Punjab Village', *Population and Development Review* 10(4): 661–78, 1984.

Rosalind P. Petchesky, 'Reproductive Freedom: Beyond "A Woman's Right to Choose"', in *Signs* 10: 92–116, 1979.

John Ratcliffe, 'Social Issues and Demographic Transition: Lessons from India's Kerala State', *International Journal of Health Services* 8(1): 123–44, 1978.

W. C. Robinson, 'High Fertility as Risk-Insurance', *Population Studies* 40(2): 289–98, 1986.

Miriam Sharma, 'Caste, Class and Gender: Women's Role in Agricultural Production in North India', *Journal of Peasant Studies* 12(5): 57–88, 1985.

Miriam Sharma and Urmila Vanjani, 'Women's Work is Never Done: Dairy "Development" and Health in the Lives of Rural Women in Rajasthan', *Economic and Political Weekly* (Bombay) 24(17) : WS 38–44, 1989.

C. Shannon Stokes, Wayne A. Schutjer and Rodolfo A. Bulatao, 'Is the Relationship between Landholding and Fertility Spurious? A Response to Cain', *Population Studies* 40(2): 305–11, 1986.

Ann Stoler, 'Class Structure and Female Autonomy in Rural Java', *Signs* (special issue) 3(1): 74–89, 1977.

Charles Tilly, 'Demographic Origins of the European Proletariat,' in

David Leslie, ed., *Proletarianization and Family Life*, Orlando, Florida: Academic Press, 1984.

Carol Vlassoff, 'The Value of Sons in an Indian Village: How Widows See It', *Population Studies* 44(1): 5–20, 1990.

M. Vlassoff, 'Labour Demand and Economic Utility of Children: A Case Study of Rural India', *Population Studies* 33(3): 415–28, 1979.

——, 'Economic Utility of Children and Fertility in Rural India', *Population Studies* 36(1): 45–59, 1982.

——, 'A Rejoinder to S. J. Datta and N. B. Nugent', *Population Studies* 38(4): 510–12, 1984.

M. Vlassoff and Carol Vlassoff, 'Old Age Security and the Utility of Children in Rural India', *Population Studies* 34(3): 487–99, 1980.

Kate Young, Carol, Wolkowitz and R. McCullagh, *Of Marriage and the Market: Women's Subordination in International Perspective*, London: CSE Books, 1981.

Sheila Zurbrigg, *Rakku's Story: Structures of Ill-Health and the Source of Change*, Madras: Sidma Offset Press Pvt. Ltd, 1984.

3

A Woman Belongs to Her Husband

Female autonomy, women's work and childbearing in Bijnor

R. JEFFERY AND P.M. JEFFERY

> Cooking bread and stew, collecting fodder and cutting it, removing cattle-dung and making dung-cakes, sweeping up ... Enough! I work the whole day—and even so my husband says, 'What do you do with yourself all day?' (Zubeida: landless Muslim Teli, separate from mother-in-law)[1]

> Village women work hard—they get no peace. If a (pregnant) woman asks for rest her mother-in-law will say, 'If you do not work how will we eat?' In villages pregnant women get neither proper food nor rest. (Village midwife)

Women are workers; women bear children. In this paper we discuss the intersection of these two sets of obligations in rural Bijnor. We consider the lives of the people most obviously affected—young married women—and how the work they do is affected by their childbearing. In doing so, we shall argue that in this part of north India differences in the households in which women live are, like class differences, a significant source of variation in women's work. A key element is the degree to which a woman's work (and the rest of her life) is controlled by her husband and her *sās* (mother-in-law). But these concerns cannot be seen only as part of a private sphere, considered without reference to public sphere concerns of production. At the most obvious level, biological reproduction (particularly numbers of sons) is crucial to issues of land fragmentation and how technological change in agriculture is associated with changes in land tenure patterns and class formation. Furthermore, the stress of multiple childbearing on women and their children has significant implications for the proponents of 'human capital' schools of development. We would therefore argue that these concerns must be placed at the centre of any account of north Indian political economy.

Agricultural technology, people's access to land and the effect of the state on production processes (e.g. through land reform or sponsoring new agricultural technologies) are clearly important, but they by no means provide an exhaustive account. Separating out a 'public' sphere, and concentrating especially on the material aspects of production, results from and perpetuates a narrow and thoroughly distorted view of the material conditions of people's existence.[2] In such an approach, a sphere of production is usually separated from a sphere of reproduction, with the first sphere misleadingly characterized by issues of class and the state (and analysed within the framework of political economy) while issues of gender are wrongly circumscribed by the boundaries of the domestic sphere (and safely relegated to feminists). But activities crucial for the day-to-day and intergenerational continuity of the social system—social reproduction, the reproduction of the labour force and biological reproduction—must be incorporated in any attempt at a comprehensive analysis, as also must be the ways people think about the social world they inhabit. A holistic approach dealing with production and reproduction at the same time is more fruitful than doggedly retaining boundaries between them.[3] Thus, women should be given their due as workers, because women are not merely engaged in reproductive activities any more than men are just involved in production.[4] Nor are women located solely in the domestic sphere and men irrelevant there. Class and gender issues pertain right across the board—and not solely because there are class differences in women's work, just as in men's. In other words, class processes and the state's intervention in them have a salience for women that has a particular character for women simply because they are women.

An individual's access to productive resources is normally mediated through households—socially constituted residential groupings, which crucially affect their livelihoods.[5] Within households, women's work differs from that of men, even when men and women alike are family workers not employees. Household members have some interests in common, and others which conflict, since men and women, young and old, have different rights in household resources.[6] Household members are recruited in two ways—by migration and birth—in both of which young women have a distinctive part. The material conditions of their existences thus differ from those of men (and from older women too).[7] In north India men inherit parental productive property (especially land, if any) and generally stay in the village where they were born, working with people known to them from childhood. By contrast, women do not normally inherit property and usually leave their birthplace on marriage, to join households where they are wives and daughters-in-law, working in the midst of strangers.[8] Such patterns reflect taken-for-granted views of proper household organization. A young married woman characteristically has little autonomy; others typically exercise control over her work, mobility and fertility. Major influences on the extent of her autonomy are the residential arrangements within which she lives in her affinal village, and the nature of her relationships with her natal kin—especially her mother. In addition to class factors, then, household politics and childbearing have a profound and distinctive impact on a young married woman, because she is both a worker and a bearer of future workers for her husband's household.

Some recent commentaries have broadened our vision of work and demonstrated that women are workers operating within the material and political constraints of their household's composition and its location in the class system. However, this entirely proper concern to resist stereotyping women as mothers does not mean that we can ignore women's childbearing roles.[9] Clearly, extending the concept of work to include reproducing the labour force can encompass the *childrearing* aspect of maternity, but pregnancy and childbirth must also be addressed head-on, rather than avoided for fear of biological reductionism.[10] Here we use the example of childbearing to highlight processes that affect a woman's capacity to influence the work she does, to move beyond the environs of her husband's house (especially as far as her own parents' house), and to influence the number and timing of the children she bears. Our aim, then, is to explore the decisions about where a woman gives birth, and in whose company, and from whom and for how long she gains relief from which aspects of her work, to relate apparently private or domestic issues to areas of life that have been more typically the preserve of political economy.

THE SETTING

On leaving the Himalayas at Hardwar, the river Ganges travels south, and Bijnor district lies on its eastern bank for the first hundred kilometres. Bijnor is similar to surrounding districts, except that it has a higher proportion of Muslims—about 30 per cent in the rural areas—and is poorer and less commercialized. Here the Gangetic plain is fertile, and where irrigation is reliable, cash-crops (especially sugar-cane) and rice are grown, in addition to the old-established staple of wheat. Most of Bijnor's 1991 population of 2.4 million people live in small villages, and subsistence and market-oriented agriculture provides their living. Our research was based in Dharmnagri and Jhakri, two adjacent villages about five kilometres north-west of Bijnor town, and five kilometres east of the bed of the Ganges.[11] Dharmnagri, established in the 1920s, had an entirely Hindu population of just over 750 in 1991. Two-thirds were peasant cultivator caste members, the largest group being Sahni, with the remaining third of the population Scheduled Caste, some three-quarters of them Chamar and the rest Jatab. Jhakri is a long established exclusively Muslim village, with a 1991 population of 475, all but 100 of whom are Sheikh. Roughly 10 per cent of the population of the two villages were in landlord or rich peasant households; 45 per cent in middle peasant households; 20 per cent in poor peasant households; and 25 per cent were landless.

In 1982–3, we lived in the Dharmnagri dispensary, on the side of the village nearest Jhakri. We made further visits in 1984, 1985, 1986 and 1990–1. Most of the material used here comes from discussions with key informants, 21 couples from Dharmnagri and 19 from Jhakri. In 1982–3, the wife was either pregnant or had recently given birth. We covered a range of castes, economic positions, household structures and parties. We worked as a team of a woman and two men, with three local women research assistants, our division of labour entailing some 'purdah of data collection'.

WOMEN'S WORK AND AUTONOMY IN NORTH INDIA

Many recent discussions of gender relationships in north India have demonstrated that women's work is under-recorded and undervalued, by officials and by husbands alike.[12] In attempting to understand the causes and the consequences (e.g. demographically) of such a situation, a number of writers have attempted to use the concept of autonomy as an organising principle to contrast the position of women in different parts of India. One prominent example is that of Dyson and Moore, who base their analysis of 'demographic regimes' on the different models of gender relations inherent in south and north Indian culture.[13] Unlike several other writers, they do not see women's 'work' (as captured, say, by the census or the National Sample Survey) or educational achievements, as fundamental.[14] For Dyson and Moore, female autonomy is the crucial intervening variable; they define it as 'capacity to manipulate one's personal environment' (p. 45). Where female autonomy is relatively high, women typically have relative freedom of movement, strong post-marital links with their natal kin, the possibility of inheriting and retaining property and some control over their own sexuality. They argue that, by comparison with areas where the 'south Indian kinship system' operates, levels of female autonomy in north India are lower, fertility and overall child mortality are higher, and instead of rough equality in female/male child mortality ratios, there is a marked excess of female child mortality over male.

Dyson and Moore recognize that economic relationships gain their meanings only through cultural and social processes. They emphasize the *control* over wages earned or property owned by household members rather than the simplistic indicator of whether or not a woman is in employment or works outside the home. Neither education nor work in themselves necessarily bring high social valuation to a woman. Moreover, Dyson and Moore stress the significance of inheritance and property-ownership: in north India, 'women generally do not inherit property for their own use, nor do they act as links through which major property rights are transferred to offspring' (p. 43). This particularly applies to productive property, especially land, but also to the mainly non-productive property which comprises the items that are most succinctly summarized as dowry, which are transferred when marriages are arranged and for many years afterwards. Young married women may have only limited rights over these items; they often act just as channels for resources (e.g. foodstuffs, clothing, occasionally cattle or loans of money or equipment) transferred from their natal kin to their marital kin. From this perspective, then, it is not so much the cost to a woman's parents that is at issue (as is often argued, for example, in discussions of son-preference and female neglect), but the degree of control over the dowry that the woman herself can exercise. Class enters into these relationships in ambiguous ways; poor women in north India may have to work for land-owning men (and women) but they may have (relative to their husbands) more substantial independent sources of income than women in wealthier neighbouring households. Even though patrilocal marriage is the norm for them as well, poor women are somewhat less subject to restraints on their mobility and controls over their sexuality and marriage; but this makes them vulnerable to sexual harassment by wealthy men. A family's growing wealth leads to increases in the restictions on its women's mobility and employment, and their increasing 'domestication' means a decline in their autonomy.[15]

In many respects, the women who are the focus of this paper are typical of those already described in the literature on north Indian women, and we will therefore merely sketch in the 'normal' patterns of their lives; we will then consider exceptions, to see what light they can throw on these patterns. Few women have been to school at all. Even fewer have acquired marketable skills. Their marriages were arranged by their parents when they were in their mid-teens or earlier, to men almost all older (and more likely to have been schooled). These features help to sustain a marital relationship in which the wife is subordinated to her husband. Post-marital relationships between a woman and her natal and her affinal kin are also highly structured, and presume that the two sets of kin are completely separate. This contrast usually also has a clear geographical component: a woman in her affinal village is under the scrutiny of her affinal kin: her behaviour in her natal village is normally very different. In Dyson and Moore's analysis of female autonomy, the north Indian woman's difficulty of access to her natal kin is seen as an important subsidiary contribution to her low level of autonomy. The more frequent the contacts between a woman and her natal kin, the more chances she has to use their support in cases of dispute with her affinal kin. Seclusion practices and the control over a woman's mobility by her affinal kin cut a woman off from her natal kin (who are presumed to have her interests more at heart than her affinal kin) and leave her to the mercy of her husband and his relatives. Locked into such a system of marriage and kinship, with control over household resources firmly in the hands of men (and, to a much lesser extent, older women), a young married woman does not have direct access to these resources; since her work may receive little recognition either, she has little autonomy.[16]

We have discussed Dyson and Moore's account in more detail elsewhere.[17] Here we will make only two further comments. Firstly, although Dyson and Moore mostly discuss women's autonomy in terms of hypothesized relationships between husbands and wives, women themselves also emphasize their relationships with their mothers-in-law, particularly in the tussles which may lead to the separation of households, as a major site where their *azādi* (freedom) is contested and must be negotiated. Secondly, in a situation where most women have very low autonomy (as defined in their terms), what (if any) factors offer women the possibility of ameliorating their positions?

In this paper, then, we want to concentrate on aspects of domestic/household organization. Women are largely embedded in the domestic units of which they are members, and family and kinship are key elements in setting the parameters of their autonomy. Limitations of space prevent us from addressing in detail the implications of differences in the class positions of the households to which young married women belong. In this paper, two sets of issues are central: sources of variation in a woman's relationships in her marital household, and as it is transformed, with other affinally-linked households in her marital village; and in her relationships with her parental home and village.

WOMEN AS DAUGHTERS AND BRIDES

> In your husband's village, shame comes. And even if you don't like keeping purdah, you still must.
> (Ghazala: middle peasant Muslim Sheikh, joint with sās)

Women's position in rural north India is generally well captured by the literature cited by Dyson and Moore and summarized above. In brief, women mainly relate to the structures of class and the state through men, and normally not directly in their own right.[18] From birth onwards, girls are treated in ways which prepare them for a married life of subordination to a husband and mother-in-law, and socially prescribed patterns of behaviour inculcate a sense of shame in women who contemplate stepping outside these norms. Here we want to draw attention to two aspects of the effects of marriage on a young woman's life: the significance—in terms of the access of young women to their natal kin—of three 'aberrant' forms of marriage; and how young women evaluate different post-marital residential arrangements.

While the bride normally goes with a dowry to her husband's house in another village, three alternative forms deviate in one way or another from this norm. They are all exceptional in the nature of the contact between a man and his wife's kin. Some grooms move to their wife's house (to become an in-living son-in-law). For men who have no sons but who own land, importing a son-in-law is a strategy to ensure some reliable assistance in the management of the land and a source of support in old age. A second pattern is that of within-village marriages, almost all Muslim; or marriages arranged within a small geographical distance. Thirdly, some marriages fall outside the pattern of a bride being 'gifted' with a dowry (however small), and she instead is 'sold' (normally by a relative other than her parents). Reactions to these forms of marriage potently reveal ideas about the sexual and domestic politics in which young brides should be situated and about the proper contacts between married women, their husbands and their natal kin. The in-living son-in-law and close marriages alike pose problems for the husband's authority; the bought bride brings neither the expected dowry nor any subsequent gift-exchanges.

The in-living son-in-law

In necessity a couple might live in the wife's village. But women have shame after they are married and stay in their husband's village. My husband would not like to live in my village. (Adesh: rich peasant, Rajput, no sās)

A man can go to his affinal village for a visit but not to live. There is shame in doing that. People would point you out to your wife's relatives. Some people do it, when the wife's parents have no sons. That way the in-living son-in-law gets his father-in-law's wealth. But otherwise a man would not like it. (Ashok, Adesh's husband; rich peasant, Rajput, joint with father and brother)

Young married women's opinions about in-living son-in-law marriages were divided. Some thought that the woman obtains peace, love and help by being close to her parents. Others said that such a marriage is not honourable: on brief visits to the wife's natal village, a couple does not cohabit because of embarrassment, but if they lived there regularly they would. Moreover, a bad husband would still beat his wife without consideration for his parents-in-law:

There would be no benefit for the woman, for the husband does what he likes. A man with no sense of shame might swear in front of her parents. Only this, the mother will show consideration to her daughter. But there is no difference for the husband. He does not keep his wife's parents happy and his wife belongs to him and must obey him. (Khurshida: middle peasant, Muslim Sheikh, no sās)

Some women considered that in everyday life there are few differences for the man who moves to his affinal village and becomes an in-living son-in-law; he can meet his parents when he likes because his wife cannot stop him, for he is unfettered (*azād*), unlike a woman. Other women, however, recognized that a man would not like the arrangement: 'A man lives with honour in his own house, but is subdued by living in someone else's.' Just as a *sāla* (wife's brother) is not respected, so a *jamāi* (son-in-law) is out of place and demeaned by living in his affinal village:

> *sās-ke-ghar jamāi kutta*
> *Bahin-ke-ghar bhāi kutta*
>
> At his mother-in-law's house a son-in-law is a
>   scrounging cur,
> At his sister's house a brother is a scrounging cur.

Certainly, men all agreed that the only benefit of being an in-living son-in-law is the attraction of inheriting more land. Being an in-living son-in-law makes a man comparable to a bride, in the sense of living among strangers, away from one's own kin. Being an in-living son-in-law undermines a man's ability to retain control over his wife as master of the house. She has constant protection from her natal kin and can move around as she cannot in her affinal village. But her husband lacks both his normal back-up from his kin and freedom from interference from his in-laws. Even on brief visits to his affinal village he must abide by the eating and sleeping times of his in-laws and behave meekly in his father-in-law's presence. He loses face, and his in-laws may not respect him as their 'guest' but challenge him over possession of the house and land. Not surprisingly, each of the key informants who had experienced life as a *ghar-jamai* had found it difficult to sustain; one temporarily gave up his effort to gain his father-in-law's land and lived in Dharmnagri for some years, though he has now returned to his affinal village. Landless people without sons have nothing with which to attract an in-living son-in-law. Just occasionally, however, a landless man moves to his wife's village for work. One such moved to Jhakri after a dispute with his stepmother and another moved to Dharmnagri because his natal village was washed away by the Ganges. These are not true in-living sons-in-law, as they neither scrounge from their parents-in-law nor support them. But the wife's parents and brothers are probably still at hand and compromise the man's capacity to exercise his authority.

Within-village or close marriage

Within-village marriage is not right. It is a shameful matter for the woman. Her parents hear everything and there will be many worries. All four parents will be pulled into fights and other relatives will be sworn at because of their connection. But there is no difference for the man as he is his own boss. (Qudsia: middle peasant Muslim Sheikh, separate from sās, married 2 kms from her natal village)

All the benefit is from a distant marriage because you avoid fights with your in-laws. Within the village or even nearby, they hear everything and you end up on bad terms over every little thing. If the marriage is distant, you can hit your wife and swear at her but no one would find out. If it's near, you can visit often and it does not cost much to travel. But if it is within-village what would happen? Your wife's people would be so used to seeing you that you wouldn't be special. They'd just say, 'Come, smoke the *huqqa*.' But if the wife's village is distant, they'd say, 'Look our guest has come from far away.' They'd fetch good food and sit you down. (Qadir, Qudsia's husband: middle peasant Muslim Sheikh, joint with father)

Only Muslims permit within-village marriages, of which there are seven in Jhakri. Yet Muslim women and men alike are often vehement in their opposition to within-village marriage. Women consider it shameful for a woman's parents to hear all her marital difficulties, and say that caring daughters would not like to worry their parents. If the woman is a bit distant from her parents, she can ensure that they never hear her 'little troubles'. Further, managing her relationships with her elders when she is a village daughter-in-law as well as a village daughter can be a problem: those people before whom she may appear one day are the very people from whom she must keep purdah the next. In addition, she can rarely stay with her parents and obtain a respite from her in-laws. Although she can see her parents often, they make no special fuss, and her husband and mother-in-law can summon her back at a moment's notice.

Men also consider within-village marriage a problem. The in-laws hear every fight and may become involved, for all that they should not meddle in their married daughter's business. Their proximity can restrain the man's capacity to rule his wife, though her relatives do not necessarily provide a refuge for her in case this offends her in-laws and threatens good relationships with them. In addition, the joys of good hospitality which a son-in-law expects are closed to the man in a within-village marriage. He becomes commonplace; the exalted son-in-law is brought down to size by familiarity.

Echoes of these concerns (in a less extreme way) are also heard when people discuss the costs and benefits of being married in a neighbouring village. Amongst Hindus, the size of their sub-caste and where it lives influences how widely they range to look for a groom. They rarely arrange marriages between people descended from a common ancestor or who are already closely connected. Their matches are always in another village, mostly with partners from over 20 kilometres away. Occasionally, marriage links are reinforced when two women (sisters, paternal cousins, or a woman and her brother's daughter) are married into the same village. However, there should be no suggestion of an exchange of women between two villages. Consequently, marriage networks among Hindus tend to be widely flung.

In Jhakri, marriages among the Sheikhs are generally arranged close by, within a small circle of families and villages. Muslims basically exclude only siblings, parents and parents' siblings as marriage partners. Marriage between close relatives is permitted, although only three of the eighty-nine ever-married women in Jhakri had married first cousins. About half the remaining marriages are between partners from villages within eight kilometres of Jhakri. Only non-Sheikh marriages are regularly arranged at distances comparable to Hindu marriages, though often between partners more closely related than Hindus would permit. Further, Muslims do not explicitly exclude exchange marriages between kin groups and villages.

Marriage distance results from decisions made by parents, but young couples themselves have to live with the consequences:

If your wife's village is nearby, you or your wife can easily visit her relatives. News of celebrations, illness and death will reach you quickly and you have time to do something. Also it is cheap to come and go. But there are disadvantages. There can be disagreements over tiny things. If you are married distantly, you live apart from disagreements. But ask my father about this: I was only the groom and I just did what I was told. (Faruq: middle peasant Muslim Sheikh, joint with father: wife Fatima from 3 kms away).

Generally, women are relatively positive about geographically close marriage. In Jhakri, women who are married within walking distance of their natal village say this permits frequent contact, through visiting by a brother or by their husband. They themselves may hurry to finish their day's work, meet their parents for a few hours and return to their affinal village to prepare the evening meal, though negotiating longer stays can be more difficult. Women in Dharmnagri and other Jhakri women more distant from their natal kin regret that important life-events can occur without their knowledge. Especially for poor people, travelling time and the cost of bus fares can cut women off from news. A woman's husband and brothers visit infrequently. She goes to her natal village only with a male chaperon to stay for some days or even weeks, but probably no more than three times a year, and less once she has been married several years. Indeed, a man often visits his wife's natal village more frequently than she does.

On the other hand, a woman married somewhat distantly from her parents is more warmly welcomed when she visits. Moreover, if her parents are close by, they are needlessly grieved by her marital troubles, in which they should not meddle. Although men grant the benefits of cheap travel and rapid conveyance of important news, their comments hinge on these two objections to close marriage:

If my wife's natal village were too close, she would get news too often—and so would her parents. There would be too much interference. It is better for a man to be married distantly. When I visit my parents-in-law they think I am good to have come from so far [30 kms] and they treat me well. I am given the best to eat. If I were married close, what would happen? They'd toss me two-three pieces of bread and ten minutes later (with luck) I might get some pickle and salt! (Tulsi: poor peasant Chamar, joint with father)

Men say that excessively frequent contact between a woman and her natal kin permits unwanted interventions, shaming her husband and inhibiting him from beating her. Being somewhat distant is better; the son-in-law is treated like a prince, not someone commonplace. Moreover, his wife must settle in her affinal village: she cannot run to her parents after every little fight. One woman came from only six kilometres away, and her husband complained:

My wife goes to her natal village too often. Her father calls her and she sometimes goes alone with the small children. I should finish this marriage and make another one from much farther away, then my wife would not be able to go so often. It would be too far for her to go alone and I would not give her the fare anyway. (Dilshad: rich peasant Muslim Sheikh, joint with father)

Marriage close by—whether within the village or in a neighbouring village—weakens the barriers that men want to maintain between the young couple and the wife's parents in order to strengthen their hands in establishing *ādmi-kā-rāj* (a husband's rule). Amongst the Sheikhs, then, dense marriage networks may offer a woman some more protection than is available to other women. Information can spread very quickly through third parties, since several women from one village may be married into another. Furthermore, other kin may rapidly become entangled in a dispute if the marriage is between a couple already related. We now consider the reverse situation—when a woman is virtually entirely cut off from her natal kin.

The Bought Bride

I bought my woman from my cousin in Bijnor. No one has ever visited her from her natal village. How could there be any coming and going or gift-exchanges? I don't know where she is from and her people do not know where she is. (Lalit: middle peasant, Sahni, joint with brother)

Several examples, including those of three of our key informants, indicate that having no parents-in-law is undesirable for a man. Men with some inadequacy—especially when not offset by wealth —may be able to marry only if they purchase a bride (*bahū mollena*).[19] The woman in such a marriage has no more control over her destiny than other married women: a go-between decides the match, receives the payment, and transfers her. The bought bride is even more shorn from her roots than the typical bride. She is unlikely to visit her natal village or receive visitors from it. One man told us that his wife had never been to her natal village since arriving in Dharmnagri:

A bought thing is bought, I gave her brother Rs. 800, so I bought her from him. He said he would treat us like other in-laws, but to this day no one has ever called us to her natal village. She sometimes asks to go but I shall not go without an invitation. Without being called there is no esteem. (Rohtash: poor peasant, Dhimar, separate from brothers)

Thus, a bought bride can expect little or no support from her natal kin. As she has nowhere to run to, her husband's authority cannot easily be undermined. She brought no dowry, however, and sustains no gift-giving: she comes empty-handed and so compares unfavourably with women in orthodox marriages. On the other hand her husband had to pay for her, and would probably have to pay to replace her: consequently he may be protective of her.

A Woman's Access to Her Natal Kin

These three forms of marriage, then, result in relationships between the new couple and the wife's natal kin which are either abnormally close (in-living son-in-laws and close marriages) or non-existent (bought brides). Men prefer a median position, which allows for a good relationship with their parents-in-law, if only to ensure that gift-giving is maintained. A daughter becomes 'someone else's property' on marriage, and her parents accept that her husband expects to control her. Yet they can pose a threat to his authority. Maintaining a balance is tricky, most feasible when marriages are arranged at some distance, a view accepted for rather different reasons by women and men. In practice, social and geographical distance is generally maintained between a woman in her affinal village and her natal kin. A wife should have kin, but, ideally, 'one's child's parents-in-law and the latrine really must be distant' (*samdhiyana or pakhana dur-hi hona chahiye*).

Marriage does not preclude contact between a young woman and her parents nor their continuing concern for her well-being. In extreme circumstances, parents may provide a refuge when their daughter's life in her affinal village becomes unbearable—as for one new bride whose husband's sexual advances had been so rough and frequent that she was terrified, or for a woman badly beaten up by her husband and sās because her dowry was insufficient. A woman's parents may refuse to return her to her husband until they are reassured about her future treatment. A married woman's father is inferior to her affinal kin, but the husband's youth somewhat counterbalances this asymmetry and he may have to accept humbling criticism from his father-in-law. Yet a woman's parents dare not resist the husband's demands for her return too strenuously, lest they precipitate future difficulties for her or (perhaps even worse) her desertion.

Even in less anguished circumstances, young married women in Dharmnagri and Jhakri were quick to contrast their life in their affinal village with their congenial existence in their parents' village. In the natal village they are loved and have warm supportive relationships with other women. Their movements and activities are not so overseen by others. They do not observe strict deference behaviour and bodily concealment: they cover their head (but not their face) out of respect for their father but can talk openly to everyone in their natal village. They find benefit (*faida*), affection (*mamta*), consideration (*khayal*), succour (*madad*) and peace (*aram*) there. But in the affinal village a young married woman is watched and must obtain permission for everything she does. She has little room for manoeuvre, and obtains respite only when she visits her parents.

There is this difference between the natal and affinal villages: in the affinal village whatever work I am responsible for I must complete. But in my natal village there is no necessity, as my mother and sisters do the work and I can work with them if I please. I cannot abandon my work in my affinal village—my husband, my sās and other women would ask why I had not done the work which there is for me to do. No one asks me that in my natal village. (Imrana: middle peasant, Muslim Sheikh, separate from sās)

For a woman there is all the difference between her own village and her husband's village. In her own village she is someone's daughter and is not responsible for her family. In her husband's village she has obligations and she has to work. (Tulsi: poor peasant, Chamar, joint with father and brothers)

A woman's visits to her natal village, however, are carefully regulated. Affinal kin are entitled to her work. Her absences cause problems for others, if her sās has to take over her responsibilities, or her husband faces food worries. Every day, her waking hours are largely occupied by her work: that in itself keeps her in her affinal village. Women often said they had no free time to take children for medical treatment in the nearby dispensary: how much harder would it be to stay with their natal kin for a while? Affinal kin may resist requests for leave, for a wife in her natal village is beyond her husband's control and her work is also lost.

> My wife cannot visit her parents without permission. She asks more often than I permit for I do not let her go if there is too much work. I have never suggested to her that she goes. When she goes away, I have problems. Why should I make difficulties for myself? (Shankar: poor peasant, Chamar, joint with father and brothers)

Bought brides have no such refuge. Similarly, women in within-village marriages cannot easily escape their obligations.

> Within the village a woman cannot rest peacefully at her mother's house. Let there be just a little work and her husband's people will call her immediately, saying, 'This work is to be done, come along!' (Ghazala: middle peasant, Muslim Sheikh, joint with sās)

Only in-living son-in-law marriage offers particular benefits: the woman's mother—while she is alive—helps with the work and gives her daughter some peace. Otherwise, an affinal village some distance from the natal village is thought best, for a woman can stay with her parents once in a while and obtain some respite from her affines. Even that, however, remains an eagerly-awaited treat. Basically, a *bahū* stays in 'her own house' in her affinal village: and while there, she should perform her own work.

RESIDENTIAL ARRANGEMENTS FOR THE BRIDE IN HER AFFINAL VILLAGE

> In my own village, when the mood is to go some-where, just go. No one will say anything. But in my husband's village I cannot go anywhere easily.(Maya: middle peasant, Sahni, separate from sās)

Once married, then, most young couples reside in the husband's village. To her husband's relatives a young married woman is a bahū (bride, daughter-in-law), and the domestic authorities in her affinal village control her movements, her activities and her contact with her natal kin. At least initially, she and her husband will probably live with her husband's parents, though her transfer to her affinal village takes place gradually. For a period of up to a year or so, the bride makes lengthy visits to her natal village. Each time she returns to her affinal village she is again accompanied by gifts of clothing and foodstuffs. Nowadays, in Dharmnagri and Jhakri, the cohabitation of most young women has taken place before they are about 17; this age has not changed much in living memory, though the gap between marriage and cohabitation has decreased from the several years which used to be typical.[20] Even after the cohabitation, the bride comes and goes between her natal and affinal villages, gradually increasing her time in her affinal village until she stays there most of the time.

In some respects, the household where the young bride lives in her affinal village is a single entity.[21] Cooking and eating together generates and reflects a commonality of interests. But the household has a political dimension, for domestic units are shot through with divisions of sex and age. The young woman's shift in residence entails the transfer of rights over her from her father to her affinal kin, especially her husband and mother-in-law. A man cannot easily monitor his wife's behaviour because he spends little time in and around the house. But within the domestic arena, women are differentiated. Those 'elder' to a married woman—her mother-in-law, or her husband's elder brothers' wives—demand respect. Through their marriages to hierarchically ranked men, women are profoundly divided from one another. The newly married woman is at the low point in her marital career. More established women are further along the track she expects to follow. They are concerned for the integrity and prospects of their own husband's household rather than to be supportive and welcoming: indeed, they wield authority over her. In practice, then, much of the normal policing of young married women is delegated to older women, especially the sās and older sisters-in-law (HeBW). For years, these women have been subordinate to the domestic authorities of their affinal village, but gradually their elders die and their growing children become subject to their authority. The arrival of a daughter-in-law over whom power can be exercised is an important transition for an older woman.

Close surveillance by the sās is not necessarily a source of overt unpleasantness: equally, however, many a young bride has unhappy tales to recount. Several older women talked of being frequently beaten by their sās, younger ones sometimes said their sās instigated wife-beating; significantly, '*susri*' (mother-in-law) is a popular term of abuse among women. Direct and close control by the sās is supplemented with a more general scrutiny by the other women in the bahū's affinal village. Her movements are regulated, her conduct commented on and her conversations observed. A bahū's female neighbours may report her misdemeanours and comment adversely if she is not beaten, saying that her husband has allowed her 'to sit on his head'. Women rarely rally round to intervene—unless the beating is excessive—and may comment that the beaten wife is only receiving her due. The husband's close female kin, then, are not ready allies, and a bahū's limited mobility inhibits the development of supportive relationships with any other women in her affinal village.

Household composition also affects the spectrum of the bahū's work, her control over it and her collaboration with other women. A bahū generally begins her marital career sharing a cooking-hearth jointly with her sās. But being joint is not necessarily a long-lived arrangement. Bahūs live in other domestic situations, separate from their sās or alone, without a sās at all. Of the 154 married women under 45 years old in Dharmnagri and Jhakri, about one-third live in each type of situation and only four share with their sister-in-law (HBW). Women normally start joint, go separate and then become alone when their sās dies. But some bahūs arrive after their sās is dead, or start their married life separate, while others stay joint until their sās dies.

Collaborative relationships among women in the affinal village do not necessarily parallel those of their menfolk. Among the landed, a man and his adult sons commonly share land and draft animals and cultivate crops together. Less commonly, a man divides his land among his sons and either retains a share of the land or levies grain and cash from his sons, or he may turn his sons out to earn their own living if they are not pulling their weight. After his death, the land and livestock are usually divided equally among the sons and they may begin cultivating separately. Sometimes, however, brothers continue to work the land jointly for many years, benefiting from economies of scale with shared assets such as tube-wells or ploughs and draught animals. Joint operations among men are less common among the poor peasants and the landless. Overall, men tend to work collaboratively for longer than their wives do. If a woman's parents-in-law are both alive, she will never continue to be joint with her sās after her husband is separate from his father. On the other hand, a man may still be joint with his father when his wife is separate from her sās. In such instances, the jointly produced income is divided and consumed separately. The women work separately and the separate cooking-hearths no longer share responsibilities for most expenditures, apart from those connected with the joint productive enterprise. A woman's experience of work is profoundly influenced by her relationship with her sās (if she has one).

Joint with the sās

I am the only bahū and I have never had a fight with my sās. She is very straightforward and if she ever criticizes me I just listen silently. But my sās does not create fights. (Najma: middle peasant, Muslim Sheikh, joint with sās)

After marriage, a bahū comes and goes between her affinal and natal homes. But her husband needs a sure source of cooked food, while her sās wants help with her work and may wish to educate the bahū in cooking, celebrating festivals, respect and obedience. Thus, the newly married son and bahū are expected to be joint with the husband's parents. Among our key informants, only one started her married life separate, because she was a second wife and her husband was already separate, and for six key informants the sās had died before they took up residence, so they began their married life alone. When a sās and bahū are joint, they share a cooking hearth to prepare food, care for jointly owned livestock, and so forth. The bahū generally does all or most of the housework. If there are animals, she probably helps to feed and water them; she may also deal with cattle dung alongside her sās. Generally, the sās and any unmarried husband's sisters do the other outside-work, collecting fodder or taking food to men in the fields if necessary. A bahū who lives with her sās, then, is unlikely to perform the full range of tasks that comprise the women's work for her household because her sās also participates. Conversely, a bahū who is joint must tolerate the authority of her sās. The work is shared, but not necessarily equitably. Often, the sās does not insist that her own daughters play an equal part and the bahū bears a heavy weight of outside-work as well as house-work. If two women remain joint, the structural difficulties of living together can best be overcome by a demure bahū and a sās who does not abuse her authority; they may co-operate peacefully for many years.

Becoming separate

If a sās has just one bahū, the two women will probably remain joint unless the men separate. But such arrangements collapse in different ways. Most commonly, the process begins when a second son is married and a new bahū arrives; the different expenses of the brothers' families aggravate the competition between them. Men resent helping to feed their brothers' children, subsidizing emergency expenses such as medical care, or meeting some of the long-term costs of marriage for a nephew or niece. These tendencies for fission lead to the creation of separate cooking-hearths while the men continue working together, an arrangement that may persist for years. But the separation of hearths begins a process that eventually results in the separation of men's work too.

Whatever other factors may have been important, disputes among women are generally given as the reason for the division of hearths when the men stay joint.[22] A bahū is said to be caught between her husband and her sās like grain being pounded between a large pestle (*masal*) and a hollowed stone in the ground used as a mortar (*okhli*). A bahū's best chance of escape is to establish a separate household. A bahū may complain to her husband about his mother's treatment and ask him to press for separation. But she can only become separate if her husband is willing, and amiable separation may be elusive. Antagonisms between a bahū and sās are often exaggerated by the presence of another bahū. The wives of brothers may fight about the sās's favouritism among her several bahūs. Their collaboration is fragile. Each woman sees her interests threatened by other women close to her, because their destinies are linked to men who are themselves competing over scarce resources. Women say they cannot trust affinal women, that there is no kindness or civility among them, that more than one woman around a hearth will certainly fight, and that women criticize one another's work.

Women cannot remain joint. They fight. One says that she has done more work, another says she has done less. That way fights begin.
Q: Don't men fight?
A: Even if one does less work than the others, there is no fight. Women do not have good morals.
Q: What if the two women are sisters, could they not work well together like brothers do?
A: There will be fighting even between sisters. Women fight more than men. (Imrana: middle peasant, Muslim Sheikh, separate from sās and two husband's brother's wives)

Two or more bahūs rarely remain joint with their sās for very long. Only 8 households out of 197 include more than two married women. Sometimes the bahū who is on best terms with the sās remains joint while the other separates or the older bahu moves out and the younger stays, perhaps until a further brother is married. The sās, indeed, may make her bahūs separate or make one bahū become separate when the next bahū arrives rather than having to live constantly with fighting.

I myself made my daughters-in-law separate at the same time. They were always fighting over their work. I live with my unmarried son and daugther and my daughters-in-law give me no help at all.

But I am better- off separate for if I had one daughter-in-law with me she would complain that she had to meet my expenses. (Viramvati's widowed mother-in-law: landless, Jatab)

Separate from the sās

When you're joint you must do the work even if you don't want to. But if you are separate, you don't have to make dung-cakes if you are ill or you can make them when you want to. That is why I have been separate in making dung-cakes for the past year, although the animals are still joint. Before that I was just separate for cooking. (Hashmi: middle peasant, Muslim Sheikh, separate from sās)

Separation may be achieved smoothly or with mutual hostility, and either in a once-for-all break or a protracted process. Moreover, separation has complex implications for the content of bahūs' work. For women in households where women's work is basically house-work, the division may be simple and quick and the bahū's workload will probably be considerably eased. In livestock-owning households, however, complete separation may take some years and entail several changes in the women's working patterns. Separate cooking may initially coexist with shared animal-work for jointly owned draught and milk animals. Later come rotas for feeding, milking and dung-work for jointly owned animals and separate animal-work for separately owned animals. Finally, the animal-work becomes entirely separate once the men have separated. Thus some bahūs acquire a more diverse workload after separation, others a more restricted one.

It is not, however, simply a question of content and quantity of work: bahūs who are separate contrast the benefits of their situation with the difficulties of being controlled by their sās and managing conflicts within the household when they were joint. Once separate, they are not necessarily free from their sās' interventions but they can say, 'What's it to do with her, now we are separate?' and they can wrest some control over their daily life in their affinal village, where self-determination is otherwise hard to achieve. The sās no longer operates such strict surveillance over her, while the bahū can more actively schedule her working-day and cook what her husband wants without regard for his relatives' preferences. Set against these benefits, however, are some costs. The bahū who is separate attains a position of greater self-determination in her work schedule, but she is now the only adult woman in her household and entirely responsible for the work. The costs of self-determination become apparent in a crisis. Women who separated under some cloud can be distinguished from those whose separation was not acrimonious. Eight of the key informants are separate but are on such bad terms with their sās that their own or a child's illness can pose problems, for they have no one to help them. Similarly, obtaining permission to visit their natal kin can be difficult. One woman separated from her sister-in-law (HeBW) and their sās within a year of marriage: her sās describes her as a slacker. She rarely goes to her natal village: as her sās commented, 'How can she go, for who is there to do her work here?'

A bahū who separated on good terms with her sās has more day-to-day independence from her sās than the bahū who is joint, but not at the cost of help in an emergency. Nonetheless visiting the natal village remains a problem. Even a willing sās can hardly be expected to remain so if the bahū makes excessive requests for leave. Bahūs often explain changes in their own visits to their natal kin and of their husband's married sisters to Dharmnagri and Jhakri with the simple refrain that once a woman is separate she cannot easily leave her affinal village. A sās, for all her other drawbacks, is still the most reliable source of help for the bahū in her affinal village. Other women have their own work and cannot be expected to rally round. A bahū may unwillingly have to submit to her sās's discipline when they are joint, and living separately may ease this for her, as she then has only her husband to deal with. But this must be offset against losing potential help with her work (either regularly or during a crisis) and the stronger ties that now bind her to her affinal village.

Being alone in the affinal village

This is my situation in my affinal village: I am alone and have to do all the work myself. But then, even bahūs with other workers in their household must work hard, for that is how things happen here. (Khurshida: middle peasant, Muslim Sheikh, no sās)

Several key informants were never joint with their sās, or were joint until their sās died. Such women describe themselves as alone (*akeli*), a term which underlines their lack of support in their affinal village. Women who are alone are solely responsible for their household, without a sās even in the wings to take over. Four key informants have a sister-in-law (HBW) in another household who might briefly be prevailed upon in an emergency. The remainder do not have even this fall-back position. Since their work is so tied to a daily routine, these women find it particularly difficult to visit their natal village often or for extended stays. Being alone roots a bahū even more firmly in her affinal village. For such women, the attractions of self-determination often seem outweighed by a sense of having no support, although if they reach this stage at the age of 35 or so, they may already have help from a daughter or bahū of their own.

Domestic cycles and secular change

Elderly women say that bahūs today are eager to separate from their sās much more quickly than in the past, that bahūs used to remain joint come what may, but now want to avoid the heavy workloads of joint living, and that now the balance of power has been reversed and the bahū 'sits on her sās's head'. One sās commented:

These days the sās is the bahū and the bahū the sās. A bahū doesn't listen to her sās. In Jhakri there are two or three bahūs who are very bad, who answer back all the time. They don't do any work. They have no manners at all. (Asghari's sās, rich peasant, Muslim Sheikh, joint with Asghari)

We cannot be entirely certain that the incidence and rapidity of separating have increased, as the older women allege. We can only speculate about the possible mechanisms of such changes. Economic and demographic changes suggest that the proportion of bahūs now separate is probably higher than previously, and the proportion alone may have declined. Thus, competition among brothers probably reflects endemic rather than novel struggles. Parents are more likely to live for longer after their children are married and, with more children surviving to maturity, the sās is likely to have more than one bahū to orchestrate. Solitary bahūs still generally remain joint with their sās in Dharmnagri and Jhakri: but a sās can rarely keep two sons and their bahūs at one hearth for long.[23] Nonetheless, bahūs may now separate more rapidly. New factors may make men more prepared to separate from their parents. The Green Revolution seems to have increased the returns from cash-crops to peasant farmers, and male labourers can probably earn a better living than before. In addition, the age at cohabitation may be slowly rising, and bahūs may arrive in their affinal village more mature and self-confident. Young women are exposed to some messages of self-assertion from the radio or from government female teachers and health staff. Even if today's bahūs are less subordinated than those of a generation ago, their freedom of action still seems, to them and to us, extremely limited.

In general, then, women separate or alone in their household have full responsibilities, while others may have more respite to offset their subordination to their sās. These contrasts are important in women's experiences of work. However, when considering control over resources, and over work and its products, differences among women retreat somewhat. Among the landed, land is vested in men, while women are dependent and controlled. Among the landless, without direct access to productive resources, women who must seek remunerated labour have to work for pittances if they can obtain work at all. Whatever their household's structure, married women are not released for full-time paid employment, nor able to use their time to establish independent means. A key feature of the position of all women remains their dependence on men.[24]

CHILDBEARING

Young married women are differentiated—if only slightly—in two main ways which could be expected to affect their autonomy: by an individual woman's own marriage and her resultant 'distance' from her natal kin; and by her relationship with her mother-in-law. We now sketch the natural history of childbearing to bring out the implications of these variations.

Pregnancy

A woman's work is lighter and optional in her natal village. The visiting daughter does not have the responsibilities of the bahū and is granted rest in her natal village, which is rarely conceded in her affinal one. But even to visit, leave alone to stay in the natal village for a protected period during pregnancy, is shameful. Embarrassment outweighs the benefits of peace and solicitous treatment in the natal village:

No matter what difficulty I faced in my affinal village, no matter if I were alone, no matter how much benefit I might get in my natal village and how much was lacking in my affinal one, I would never go to my natal village after six complete months of pregnancy. (Durgi; middle peasant, Sahni, separate from sās)

Remaining distant from their natal kin can present acute problems for pregnant women living in their natal village. Indeed, such women's problems are yet another objection to in-living son-in-law and within-village marriages, and these women are not especially privileged or able to benefit greatly from the proximity of their parents. With in-living son-in-law marriages, the issue is usually resolved by sending the woman to her affinal village in late pregnancy, where she must participate in the work there. Muslim women in within-village marriages are necessarily close to their natal kin and cannot be removed to a distant affinal village, but they curtail their contacts with their natal kin once their pregnancy is apparent. The bought bride normally has no contact with her natal kin, so pregnancy brings no special disadvantages for her. For the vast majority, however, the experience of pregnancy—especially a first pregnancy—heightens the sense of being cut off from the natal kin. Only in a few cases of severe illness did a pregnant woman benefit from the rest afforded in the natal village, while women marooned with their parents after marital rows said they were so embarrassed that they could not enjoy the release from their obligations.

Thus, pregnant women usually have to work in their affinal village, although heavy lifting work can cause a baby to fall, evil spirits may attack pregnant women wandering outside domestic space, and displaying the 'drum strapped to her belly' is shameful. The structural position of bahūs in their affinal village is reflected in common features of their work when pregnant. A bahū has responsibilities, and should not be a burden on her in-laws. Antenatal maternity leave is rarely considered appropriate, and most women were disdainful when we asked if their work had reduced during pregnancy. A household's economic requirements cannot easily be met if women relinquish their work, even temporarily. Pregnant women may adjust their activities, sitting down to work or working in snatches, carrying lighter head-loads of dung and dumping it in the midden instead of crouching to make dung-cakes. Some, too, are relived of most outside-work—but others carry head-loads of fodder or water livestock even in early labour. Our informants' working days during pregnancy are minimally 12 hours and often more, just as at other times. Pregnant women generally continue with their normal duties and move around their affinal village much as usual.

The details of pregnant women's work vary, for a woman's work more closely relates to seasonal factors and to the class position and composition of her household than to the stage her pregnancy has reached. One woman, for instance, had heavier work late in pregnancy than during other seasons. In April, when she was over six months pregnant, she had to carry head-loads of fodder with her toddler son (who had chickenpox and could not walk far) astride her hips:

These days I collect all the fodder, for the men have been too busy in the fields cutting wheat, harrowing the earth and laying new sugar-cane. Men collect fodder when it's raining or when they are not so busy—at such times I get some peace. Otherwise I must do it, even if I am feeling unwell. Also, the weather is getting hotter, so I must take drinking water to my husband in the fields several times a day. (Maqsudi: middle peasant, Muslim Sheikh, separate from her sās)

Significantly, she was separate from her sās and husband's unmarried sisters, who are disinclined to help except with the joint animal-work. Three-quarters of the key informants were either separate or alone during their last pregnancy and had no help with their work. When pregnant, such women have little option but to continue performing all their usual tasks. A pregnant woman has her greatest chance of respite from her duties if she is joint, which she is most likely to be early in her childbearing career. Almost all our key informants were joint during their first pregnancy and several had a number of children while still joint. Barring those who became ill, no bahū who was separate reported being helped by her sās during pregnancy. By contrast, all the bahūs still joint were relieved of much (if not all) of their outside-work late in pregnancy. But being joint does not entitle the bahū to shirk even during pregnancy, since 'it does not look nice for a bahū to rest while her sās is hard at work'.

How can a pregnant woman get peace? If she is joint, she does her work because she feels she should work alongside her sās. And if she is separate, she has enough work of her own to prevent her getting peace! (Urmila: poor peasant, Sahni, separate from sās)

Before the delivery, a husband's married sister is not usually called (as she may be afterwards)—and even so, she may not be much help to the bahū.

My sister-in-law stayed for a month before I gave birth but I had scarcely any rest. I don't have so much power over anyone in my husband's house that I can cause them to work. (Jamila: middle peasant, Muslim Sheikh, no sās)

Within the affinal village itself, even a husband's unmarried sister and the sās cannot be guaranteed to help unless they are already sharing work responsibilities. A pregnant woman is not polluted, and poses no danger to other people. The desirability of some rest during pregnancy (because of the *sharm* of displaying pregnancy and possible threats to foetal well-being) is in practice heavily outweighed by the requirement that the bahū fulfil her responsibilities without disrupting other women's agendas. Pregnant women in Dharmnagri and Jhakri can usually obtain only negligible respites from their work, reflecting the low levels of everyday co-operation among women.

Delivery

Pregnant women do not negotiate where they deliver or even consider it debatable. Giving birth in the affinal home is absolutely taken for granted by all women, whether rich or poor, and is what the overwhelming majority of women in rural Bijnor do.[25] Some had never heard of any alternative. A baby should be born where the 'work' (i.e. sexual intercourse) took place. Going to the natal village to deliver is anathema—even if, as women acknowledged, more ease and better care are available there. Comfort is outweighed by embarrassment.

If I am too ashamed to visit my natal village when I am pregnant, don't you think it would be even worse to give birth there? (Imrana: middle peasant, Muslim Sheikh, separate from sās)

Giving birth is a defiling event and during labour a woman becomes intensely polluting. Moreover, despite covering herself, she undergoes the shame of being touched and exposed. Women in in-living son-in-law marriages were generally taken to their affinal village late in pregnancy. Even Muslim women in within-village marriages who cannot deliver far away from their natal kin cite shame as grounds for avoiding their mothers. The baby should be born 'in the house of those people to whom it belongs', in the paternal grandparents' place.

Babies are always born in the husband's house. What, would it look good to give birth in my parent's house, in front of my father and brothers? (Nirmala: middle peasant, Jatab, separate from sās)

Overall, we heard of only a handful of women who had ever given birth in their parents' house. One in Jhakri was married to a man whose first wife had died in childbirth; in Dharmnagri, another's previous child had been stillborn. For both, the shamefulness was counterbalanced by fear of further misfortune.

Female natal kin, especially the mother and sister, are also excluded from the delivery by considerations of shame. Generally, mothers and daughters alike regard this as only proper. A woman's mother or sister is not called from another village, and probably will not know about the labour.

My mother came to visit. But she went away when she heard that I was having pains, saying that she should not be there at such a time. (Sabra: poor peasant, Muslim Sheikh, separate from sās)

Unmarried girls are excluded, as they should not learn about such shameful matters before marriage, so the delivery is attended by married women from the labouring woman's own affinal household and compound, and from neighbouring ones. The sās is central, whether she is joint or separate. If the sās is dead, her position is probably taken by the labouring woman's sister-in-law (HBW) or her husband's aunt (HFBW).

Unusual residence patterns create awkward problems and exceptional solutions. A woman in an in-living son-in-law marriage might deliver in her parents' house, but some women said that their mother and sisters would remain outside. Others considered the shame would have evaporated long since and the mother and married sisters could attend the delivery. In practice, though, most such women give birth in their affinal village. Muslim women were unanimous that women in within-village marriages should deliver with their female affinal kin in attendance, not their mother or sisters. The handful of cases when a labouring woman's mother was present were all deliveries of women in within-village marriages when the woman's life was thought to be endangered.

The post-partum period

A new mother is unclean for five weeks. For all that time no one should eat food which she has cooked. But if there is no one to cook for the family, she may start cooking for them out of necessity after a fortnight. If she is completely on her own, her husband would cook his own meals. (Village midwife)

Being a *jachā* (newly-delivered mother) is one of the rare occasions when a bahū is entitled to relief from her normal work. Her role as new mother ought to put her role as worker in abeyance and she should be temporarily replaced by another worker. Preparing food is work common to bahūs in their affinal village. But, while a jachā may cook for herself, she should not cook for others. Childbirth pollution endangers their well-being, though informants were vague about what symptoms might ensue. Further, outside-work—the dung-work, the animal-work, or taking food to men in the fields—is compromised because of the jachā's perceived weakness and vulnerability. Ideally, the jachā obtains help with at least some of her work for five weeks.

Alongside these aspects of maternity is one other that has crucial implications for the help a jachā may receive: the jachā's condition is embarrassing, her physiological processes are shameful, distasteful and striking evidence of her sexuality. Thus her access to her natal kin is restricted. They do not attend any celebrations that may take place in her affinal village and may even not learn that she has given birth for some weeks. If a bahū's natal kin know about her pregnancy, her brothers and (especially) her father may avoid visiting her around the time the delivery is expected. If they call shortly afterwards, they stay only briefly, for their presence is inappropriate. The jachā is most unlikely to be visited by her mother or sisters soon after delivery, unless she or the baby are seriously ill and this stricture is waived. And her female natal kin are not considered suitable replacements to do the work from which she is temporarily relieved. This is so even for Muslim jachās in within-village marriages and women in in-living son-in-law marriages who have been sent to their affinal villages to deliver.

You get ashamed in front of your mother. That is why she is not called after her daughter has a baby. It is an affinal matter, and it is correct that it is dealt with by affines. (Sabra: poor peasant, Muslim Sheikh, separate from sās)

A woman's natal kin may be called upon only in exceptional circumstances which need elaborate justification. Their involvement is shameful, 'worse even than involving them in a marital row'. It would also leave the jachā open to accusations that she is trying to evade her obligations to give gifts to her affinal kin. Several women said a sister's help would be congenial, but unthinkable unless they could find no one else to help.

Further, a visit to the natal village is not permissible for the jachā for at least five weeks. Solicitous care, better food, respite from responsibilities or access to medical help might all be available there. But the embarrassment of being a jachā is compounded by her defilement and her own and the baby's vulnerability, which make travelling inappropriate. Only after a final cleaning bath can she take the baby to her natal village, and stay some weeks. Such visits are not entitlements, however, and many of our key informants did not make them. In any case, the timing of such visits does not provide for recovery from the delivery itself.

Thus, the jachā whose parents are still living, whether distantly or close-by, is effectively no better placed than bought brides or women whose parents are dead. In the immediate post-partum period, the typical jachā has no access to the women with whom she can relax or the place where she is most at ease. She is tied to her affinal village and must regain her strength and her normal condition there, subject to the restraints and responsibilities of a bahū. All her most suitable female helpers are affinal kin and resorted to in preference to her natal kin, particularly her own mother or sisters.

Women say that the jachā should be helped by her sās, her husband's sister (or even both) or some other woman from her affinal connection.

There could be benefits for a woman if she stays in her natal village to give birth. But I am fine in my affinal village as my husband's married sister helps with my work. I am more fortunate than many women, who get no help in their affinal village. (Bhagirthi: rich peasant, Rajput, no sās)

Women's experiences of the post-partum period are so diverse, however, that they cannot be captured succinctly. Different women report an array of experiences and there is diversity even in individual women's childbearing careers. The support a jachā obtains with her different types of work, from whom, and for how long are influenced by several parameters which intermingle in different and sometimes unpredictable ways after each birth. Most salient are the composition of the jachā's household and her normal patterns of co-operation in her affinal village; the availability of her husband's married sisters to help; changes over time; class factors (principally in relation to differences in women's workloads); and ethnicity. Briefly, women with the heaviest workloads, particularly those in middle-peasant households, tend to obtain the least respite from their responsibilities, especially if they are Muslim.

The jachā can expect help only from women of her own household or of closely related ones, especially her sās. But the mere existence of the sās is no guarantee that she will take over the jachā's work, for prior decisions about household organisation, particularly about remaining joint or becoming separate, are important in this. Almost one-quarter of the key informants were still joint at their last delivery and they all had their work taken over by their sās. By contrast, half the key informants were separate and only half of them were helped by their sās. The remaining quarter of the key informants were alone. Other married women, living in houses sharing the same courtyard or beyond, have 'their own work' and would rarely be asked to help. Even the jachā's sisters-in-law (HBW) are unlikely to offer their services, unless (as is rarely the case) they are joint with the jachā. The sās is the jachā's most likely helper. Of the key informants with a sās still alive, only about two-thirds, however, were helped by her after their last delivery.

The jachā's *nand* (husband's sister) is a central figure in the celebrations after a birth and is also regarded as a potential helper with at least some of the jachā's work. Virtually all the key informants have at least one nand, sometimes, of course, shared with their sisters-in-law (HBW). About one-quarter of the nands are still unmarried, sharing a cooking-hearth with the jachā or in a separate one with the jachā's sās. An unmarried nand aged ten or so will probably help the jachā, just as she might on a day-to-day basis. But three-quarters of the key informants' nands are married and so probably living in another village. Help from that source is less certain, as several considerations affect whether the married nand will be called after the jachā gives birth.

The jachā's own household composition is important in negotiations over calling the married nand, for if the jachā's sās is still alive, she, not the jachā herself, is key in the decision. The initiative rests with the sās to ask one of the men to leave their work and fetch the nand from her affinal village. Once alone, the jachā herself can ask her husband to call a helper, and over half the key informants in this position obtained help from a nand. The situation of the jachā who is separate is much harder to characterize. Being separate is generally symptomatic of some distance between the jachā and her sās. The sās herself is a less reliable source of help, and, moreover, is less likely to call the nand. The jachā may face difficulties negotiating who is called to help her.

In any case, even when a nand is called after her brother's wife has given birth, her expectations of relief from her own responsibilities in her husband's village may be dissonant from those of the jachā, who (unusually) is entitled to rest herself:

My nands do not work when they come here. Their mother does it all. That is why she has not called my nands except after my first baby. She says they do no work so what is the benefit? My sās does everything, even the work for the nands' children! So I just give them gifts to mark the birth some other time when they come to Dharmnagri. (Maya: middle peasant, Sahni, separate from sās)

The nand can also come only if her affines are prepared to send her. Like the jachā, she is a bahū and subject to limitations on her activities. The most common restraint is the nand's own pregnancy or recent delivery, since visiting her natal village then would be shameful; her natal kin would not even bother calling her if they knew her condition. Sometimes, the nand returns only briefly to her natal village, if at all. The accounts given are legion, and primarily hinge on the precise circumstances in which the nand is living, on her responsibilities and the composition of her affinal household. Six of the eighteen key informants who were helped by their nand after their last deliveries received that help rather fortuitously.

Generally, the men of the household are minimally affected by the jachā's inability to perform her work. They might adjust their activities slightly or do extra work to alleviate her work if the jachā is separate or alone and the men's work is slack. Occasionally, men collect and chop fodder without help from the women. Sometimes, they eat before going to the fields or take food with them. A jachā would not be expected to do field-work. But hardly any men helped their wives with 'women's work', never with dung-work and rarely with house-work. The three bought brides among the key informants were all somewhat unusual in this. One was alone and her nand was not called. Her sister-in-law (HyBW) did the house-work and dung-work for four days and her husband collected and chopped the fodder. The two other husbands both did some of the house-work. One of them was so anxious about his wife's recovery from her caesarian section that he cared for their daughter and did the cooking for over a month. His wife cared for the baby boy, while neither her sās nor her nand offered any assistance (though both had taken turns attending her in hospital).

Another angle on jachās' experiences can be obtained by tracing women's childbearing careers, for the domestic circumstances of the jachā and her married nands change, and later births generate less celebration. Most bahūs begin their married life joint, and the sās is expected to keep her bahū under her wing until at least the first child is born. Thus, the jachā is usually best-placed after her first delivery. Almost all the key informants had help from their sās and at least one nand after their first delivery. Gradually, configurations change. The jachā who is separate is less certain of being helped by her sās. Over the years, the sās may become frail or die. Unmarried nands are converted into married, and their availability to help the jachā is compromised by their affinal responsibilities. The nand cannot come repeatedly. Only in the early years after marriage (if at all) are married nands joint with their sās or sisters-in-law (HBW). Later, the nands are likely to be separate or alone and their own successive pregnancies and growing responsibilities make it increasingly difficult for them to leave their affinal village for long. No matter what a nand's responsibilities or how many brothers she has, however, she will probably make every effort to come to her natal village after the first deliveries of all her brothers' wives, for then the celebration and gift-giving are greatest. After later births, the jachā's sās may refrain from calling the nand. This general progression is only likely to be reversed temporarily if a woman has a son after a series of girls.

The help that a woman receives can also be analysed by considering who helps with what kinds of work, involving some differentiation by class and by ethnicity. Class is significant because the length of help a jachā obtains depends in part on her workload and on the leeway which her potential helpers have with theirs. Middle-peasant women tend to have the most outside-work: taking food to the fields, watering and feeding several animals, collecting fodder and making dung-cakes. Unlike rich peasants, they have no help from farm servants, nor can they easily afford to find and reward helpers whose work can be done by someone else. Unlike most poor peasant and landless women, middle-peasant women have cattle, and are mainly responsible for their care. A middle-peasant woman's nand is also likely to have animals and other outside work for which she is responsible, and thus to find it hard to get away from her affinal home. As a result, the women with the heaviest day-to-day work obligations also tend to be the women who receive the shortest assistance—rarely more than two weeks with cooking, and for a few more days with their outside work.

The most striking differences in the quality of help received, however, suggest that ethnicity is another important consideration. Because of the density of marriage networks among Muslims, we had expected Muslim jachās to have more comprehensive support from other women than Hindu jachās. Over half the Muslim key informants with married nands had at least one within easy walking distance. Women married close to their natal home have greater difficulty in obtaining permission for lengthy stays there, but Muslim nands are also less likely to be summoned, which suggests that Muslims tend to regard the provision of a helper as less imperative than do Hindus. Among Hindus, the contrasts between the celebration following the births of boys and those of girls are more marked than among Muslims, and we had expected that Hindu jachās giving birth to boys would obtain correspondingly more help than those giving birth to girls, and that the differences would be more marked than for Muslim jachās. But the baby's sex seems to make little difference to the length of time a Hindu jachā received help, whereas Muslim jachās who gave birth to boys generally had more help than those who gave birth to girls. Even so, four Muslim jachās who gave birth to boys had fewer than five days help with their cooking.

Basically, Muslim jachās often obtain only half as much help with their house-work as do comparable Hindu jachās. For example, Hindu women do not cook for others for at least a fortnight, but over half of the Muslim jachās had less than a week off. The ethnic contrast is especially clear-cut among jachās in middle-peasant households, particularly if the jachā's outside-work is taken into account. Some Hindu women had two weeks off both cooking and outside-work. Others did no outside-work for much longer. But among the Muslim jachās, virtually all were doing outside-work less than ten days after delivery: four of the key informants were doing dung-work within a week of giving birth. Only two Muslim middle-peasant women had more than a fortnight's help with their outside-work. The differences are most marked for those with several children who give birth to a girl, and whose work is associated with land—or livestock-ownership, but they show up in almost all comparisons. Ethnic differences in the length of help which jachās obtain from any source can best be understood by noting that for Hindus, childbirth pollution is far more important than for Muslims. Although women in Jhakri and Dharmnagri alike talk of five weeks as the time during which the jachā is defiling, Muslim jachās generally begin to cook much sooner after delivery than do Hindus.

While the jachā's condition warrants some special arrangements, then, the details vary widely in practice and the perceived needs of the jachā are far from universally met. Women living amid so many others might be expected to obtain abundant help and support after delivery, but the demands of a labour-intensive economy and the low levels of co-operation among women affect the rest that jachās obtain. Help must be negotiated, and, as a bahū in her affinal village, the jachā is ill-placed to command anyone's services, and has no guarantee that others will do so on her behalf. Indeed, the weight of other considerations may be incompatible with providing lengthy help for her. Jachās in Dharmnagri and Jhakri are surrounded by other women, all of whom are workers themselves and the availability of help for more than a few days after delivery, from within the jachā's affinal village or from elsewhere (especially from a married nand), is often problematic.

The toll of childbearing

Bringing up children rests on me, not my husband. And now my own health is not recovering. The more children, the more my health sinks. (Dilruba: rich peasant, Muslim Sheikh, joint with sās; sixth child born in 1984)

Women these days have a baby nearly every year, so how can they keep their strength? (Rajballa's sās; poor peasant, Dhimar)

Young women in Dharmnagri and Jhakri consider that the domiciliary post-natal care available to them cannot ensure recuperation from pregnancy and childbirth. It is part of a more general syndrome that operates in relation to childbearing women, to the detriment (women say) of their own and their children's well-being.[26] For most women, each pregnancy, delivery and post-partum period is not an isolated episode but one part of a sequence. Moreover, women are vocal about the strain of combining their work (including childcare) with such childbearing careers. They consider that the physical demands of repeated childbearing damage a woman's health and that prolonged breast-feeding plays a considerable part in this. Almost half the key informants were breast-feeding during their last pregnancy. Generally, they had weaned their previous child by the fifth month of pregnancy, though a handful of Hindu women suckled up to the last month and even into labour. Most of the remainder had not been in a position to breast-feed while pregnant, because it was their first pregnancy or because their previous child had been stillborn or had died. The local view is that to continue breast-feeding during pregnancy also takes its toll on women's health and is not good for the suckling child. Yet breast-feeding and pregnancy tend to shade into one another, and few of the women approaching or past menopause have had much time free from pregnancy or lactation during their fertile years.

Furthermore, it is hard for a woman to escape her responsibilities in her affinal village, even if she or her children are sick. The work must be done; how else can the rest of the family survive? Going for medical treatment (especially a trip to Bijnor escorted by a man) is costly in terms of adult workers' time. Chest complaints and colds, fevers and diarrhoea, septic cuts, boils and conjunctivitis, mean that babies and toddlers are frequently fretting and take their mother's attention away from their other work. But seeking medical treatment is often delayed until the condition is serious. Paying attention to women's and children's health is often squeezed to the margins. Moreover, women cannot easily take unilateral decisions about medical expenditures. Decisions may have to wait until their husband returns home in the evening. Occasionally (as discussions of some child-deaths indicated) that was simply not soon enough. In any case, even rich peasants do not always have ready cash; poor women face even greater problems. But it is not simply a matter of time and money. On several occasions a woman's parents took her back to her natal village for medical treatment and convalescence (not only after childbirth) on the grounds that neither would be adequately granted in the affinal village. Parents, it seems, are much less likely to grudge time and money on their married daughters.

Thus after childbirth, many women receive less help than they consider reasonable and their due, let alone than they might consider ideal. These considerations are exacerbated for young married women in the affinal village by their restricted access to cash, their time-consuming work and by restraints on their mobility, all circumstances not of their own making. In addition, young women are in a very different situation from their husbands in relation to childbearing and childrearing. They experience costs of repeated childbearing which do not directly affect men. Death in pregnancy or delivery is a common fear, their own loss of 'spirit' and the anxieties of childrearing are frequent refrains. Success in rearing robust children is achieved only against heavy odds, yet children (especially, of course, sons) are necessary for the future security of women themselves and of their affinal household. In such a context, young women approach family building with deep-seated ambivalence.

CONCLUSION

What, then, are the implications of our discussion of young married women's work and how their contribution is replaced at the time of childbearing? Firstly, we want to stress that there are variations in women's positions. Class differences matter, most directly via women's normal workloads and the differing degrees to which women's activities are located within the domestic arena or outside it. Furthermore, the purchasing power of a woman's affinal household affects the general lifestyle of its members, while the affluence of her natal household affects its ability to know about its daughter's married life and to enhance her status and security through gift-giving and fulfilling obligations when additional expenses have to be met. Differences in household composition—in terms of a woman's relationship with her sās, as well as her husband's relationship with his father—make material differences to a woman's everyday life. And, although Muslim women get less help after childbirth, in other cases of adversity the differing patterns of marriage networks might mean that Muslim women can call on more support from their natal kin *in extremis* than can Hindu women married at greater social and genealogical distances.

Nonetheless, we would argue that these differences are not sufficient to counterbalance the effects on most women of their enforced migration away from their natal village on marriage, their lack of control over substantial economic resources, the controls over their mobility and their work, and the general powerlessness of young married women in their affinal household. Despite their common lack of power, however, young married women show few signs of solidarity or unity. Households are separated from one another. The familial ideology, and the norm that women should do their 'own' work, means that contacts that might subvert their subordination are inhibited. Caste and ethnicity similarly undermine the possibility of building up supportive networks. But in any case, households have diverse and often competing interests: a woman's security is far too bound up with the well-being of the household of which she is a member for major material benefits from mobilising with other women to be apparent. A young married woman looks forward to being able to resolve her problems through normal life-cycle changes: as she gets older, the authority of her sās will diminish and she will become a sās herself with bahūs to manage. Any commitment to bettering the status of the young married woman will thus be undercut by her knowledge that she in her turn will benefit from the system.

Our analysis also relates to some of the debates in the narrowly defined realm of political economy. For example, most discussions of the changing nature of labour relationships under the impact of the 'Green Revolution' focus on labour relationships outside the household. But labour relationships do not just operate in the 'public sphere' or in the context of employment (whether as bonded labourers or wage labourers). Labour relationships also occur inside the household, and issues of power and control over the labour and the resources one he produces are pertinent there too. We need to ask how changes in labour relationships *inside* the family respond to changes in labour relationships outside it, and vice versa: as yet, we are only beginning to scratch the surface of understanding what is going on.

The relationships between land, labour, and material resources are also important. Childbearing is a cumulative activity, adding up to a maternity career; and fertility and mortality—especially child mortality—are changing. Women seem to have little control over this, but many would like to reduce the number of children they bear. Although they emphasize health problems for themselves and their children, and the difficulties of juggling child care with other work, they also—like their husbands—point to the costs of children, in terms of the implications of having many children for the size of landholdings and the pauperization which can come from land fragmentation, where few alternative employment opportunities are being generated. They also feel, often more strongly than their men, the impact of growing population pressure on common property resources as soils are degraded and farmers encroach on 'unproductive' jungle land to produce more grains for human consumption. How couples reach decisions to control fertility (or not) has profound significance not just for the couples themselves, but also for long-term economic development and class formation. Thus, marriage, household structures, women's visiting arrangements, their work, their childbearing and their autonomy are not solely the concern of sociologists and anthropologists; they also have a legitimate and vital place in the work of political economists.

NOTES

1. This paper draws heavily on our recent book, P.M. Jeffery et al., *Labour Pains and Labour Power*, London: Zed, 1989. For more detail on our selection of respondents, the sources of quotes, allocation of respondents to classes, etc., see pp. 226–37.
2. See, for example, F. Edholm et al., 'Conceptualising Women', *Critique of Anthropology*, 3(9–10): 101–30, 1977. C. von Werlhof, 'Notes on the relation between sexuality and economy', *Review*, 4: 32–42, 1980.

J. Sayers *et al.*, eds, *Engels Revisited: New Feminist Essays*, London: Tavistock, 1987.

3. V. Beechey, 'On Patriarchy', *Feminist Review*, 3: 66–82; G. Omvedt '"Patriarchy:" The Analysis of Women's Oppression', *Insurgent Sociologist*, 13: 30–50, 1986.
4. L. Beneria, ed., *Women and Development: The Sexual Division of Labour in Rural Societies*, New York: Praeger, 1982.
5. L. Beneria and G. Sen, 'Accumulation, Reproduction and Women's Role in Economic Development: Boserup Revisited', *Signs*, 7(2): 279–98, 1981; B. Agarwal 'Women, Poverty and Agricultural Growth in India', *Journal of Peasant Studies*, 13: 165–220, 1986.
6. K. Young, *et al.*, eds., *Of Marriage and the Market*, London, CSE Books, 1981; N. Folbre, 'Hearts and Spades: Paradigms of Household Economics', *World Development*, 14: 245–55, 1986.
7. M. Cain *et al.*, 'Class, Patriarchy and Women's Work in Bangladesh' *Population and Development Review*, 5(3): 405–38, 1979; U. Sharma, *Women Work and Property in North-West India*, London: Tavistock, 1980.
8. M. Marriott, in M.N. Srinivas, ed., *India's Villages*, Calcutta: West Bengal Government Press, 1955, gives a man's perspective on this shift in residence: 'one's daughter and sister at marriage becomes the helpless possession of an alien kin group' (p.101). Because women move, and men stay, as U. Sharma ('Male Bias in Anthropology', *South Asia Research*, 1(2): 34–8, 1981) points out, village-based accounts of north Indian kinship inevitably take a man's perspective.
9. A. Phillips, *Divided Loyalties: Dilemmas of Sex and Class*, London: Virago, 1987.
10. For more details on this point, see P.M. Jeffery *et al.*, *op.cit.*, 9–68.
11. For more details on the villages and our research strategy see P.M. Jeffery *et al.*, ibid., especially Chapter 2 and Appendices 1–3. Andrew Lyon was the third member of our research team, and we are very grateful for his contribution to the research and to the analysis (particularly of men's attitudes, and issues concerning fertility) which was an essential part of the work which has gone into this paper.
12. K. Young, *et al.*, eds., *op.cit.*; B. Rogers, *The Domestication of Women: Discrimination in Developing Societies*, London: Kegan Paul, 1980.

13. T. Dyson and M. Moore, 'On Kinship Structure, Female Autonomy and Demographic Behaviour in India', *Population and Development Review*, 9(1): 35–60, 1983.
14. K. Mason, *The Status of Women: A Review of its Relationships to Fertility and Mortality*, New York: Rockefeller Foundation, 1984; Z. Sathar *et al.*, 'Women's status and fertility change in Pakistan', *Population and Development Review*, 14(3): 415–32, 1988.
15. B. Rogers, 1980, ibid.
16. We should note, however, that some husbands hand over to their wives all their incomes; but women deny that this managerial role gives them any independence.
17. R. Jeffery, P.M. Jeffery and A. Lyon, 'When did you last see your mother?', in J. Caldwell *et al.*, ed., *Advances in Micro-Demography*, London: Kegan Paul International, 1988.
18. U. Sharma, 1980, ibid.
19. Note that the brothers of these men had orthodox dowry marriages; there is no evidence of bride-wealth becoming common; nor of it having been so in the recent past.
20. Some scepticism must be attached to these estimates: most women had only very vague notions of their ages at marriage and cohabitation.
21. We are not here discussing the complicated relationships between a household as a production unit, which may differ from a consumption unit or common cooking hearth (called in Hindi, *chulha*); nor the differentiation of interests within these units.
22. D.G. Mandelbaum, 'Sex roles and gender relations in north India', *Economic and Political Weekly*, 21(46): 1999–2004, summarizes a wide body of literature by noting that the blame for break-up is 'usually put on the wives'.
23. P. Kolenda, *Marriage and Household in India*, Jaipur: Sterling Publishers, 1987.
24. Landless widows are the only partial exceptions to this generalization; if they have no son prepared to support them they have to find what work they can.
25. Only 3 per cent of deliveries reported by any of the women took place in a hospital, clinic or in the woman's natal village. The exclusion of parturient women from their natal home is common in western Uttar Pradesh, but unusual to the west (in Haryana and

Punjab), to the south (in south-west U.P. and Madhya Pradesh) and in central and eastern U.P.

26. Here we are not concerned with western medical views of desirable behaviour, but women's own views of their entitlements and how far they are met in practice.

4

Analysing the Reproduction of Human Beings and Social Formations, with Indian Regional Examples over the Last Century

ALICE W. CLARK

INTRODUCTION

The history of women in India has been obscured by their extreme privatization.[1] One method for unveiling the story of women's lives is to examine India's demographic history, since roughly three quarters of all vital events (all births, and half of all deaths) occur to females. Out of an interest in women's lives, I have been involved in studies of India's demographic and more precisely its reproductive history.

Thus, reproduction is my theme here; but immediately, qualifications are in order. What meanings are encompassed within this term? I shall begin by exploring a number of them, considering how they relate to an analysis of gender. Then, proceeding to some specific observations on the demographic history of India, I will show how that history—intensely gendered as it is—can be illuminated by attention to *more than one* of these strains of thinking around the term reproduction.

Reproduction is a sphere of activity the understanding of which seems to speak centrally to the analysis of the status and position of women, and to the question of how gender and political economy interact. Within the framework of a considerable Marxist feminist literature that has developed, analysing reproduction has become a central problematic for this reason. In much of this literature, in order to explain the relations of gender—and the position of women—from a political-economic standpoint, the reproduction of both life and society are considered, at least theoretically.

The Marxist feminist discussion has tended to emphasize universal features of the relationship between women and political-economic structures. Much effort has been made to distinguish between biological reproduction and what is called social reproduction. An especially thorough discussion appears in Edholm, Harris and Young (1977). These writers separate for analysis three spheres of reproduction, broadly defined: social reproduction, the reproduction of the labour force, and biological reproduction. They acknowledge that two or more of these will often completely merge or overlap, but urge that they be conceptually separated. The distinctions as well as the interrelationships among them are important in explaining the position of women in a society, they argue.

I will attempt here to trace the intellectual background of several aspects of what is called *social reproduction*. While the concept derives from Marxist thinking, in some of its aspects it has migrated into feminist theorizing which may or may not be essentially Marxian. This review is not meant to be systematic, but rather is meant to suggest which lines of thought have led to helpful insights in an analysis of gender in relation to political economy.

I. THE CONCEPT OF SOCIAL REPRODUCTION

A. *Reproduction of the structure of society*

Social reproduction is a term used in the first instance in Marxist literature to refer to the overall reproduction of social formations and, in particular, of the mode of production. In this context, it refers to the reproduction of those economic structures believed to define and determine the underlying processes within society at the broadest level. Determination of society by the structure of its economy is a basic postulate in major Marxist traditions.

In this formal sense, as reproduction of the material structure, social reproduction is abstract enough so that there can appear to be some question as to whether it needs to entail any analysis of gender. To discuss social reproduction at this level may seem only to require attention to the gender issue when contradictions occur between the forces and the relations of production, in such a case where gender (as a relation of production) is involved in the contradiction. At this level of abstraction, discussions of social reproduction may be opaque on the subject of gender, and often are.

Discussions of gender in a political-economic context, however, cannot be so opaque on the subject of the reproduction of social formations. The conditions within which the material structure is reproduced necessarily incorporate biological reproduction, and this cannot be discussed without analysing and specifying the workings of the system of gender. Marxist feminists have been aware of this, while other Marxist scholars have not attended to it consistently or adequately.

B. *Social reproduction and social ideology*

Another part of the Marxist canon has linked the notion of social reproduction with ideological processes. A shift within Marxist thinking toward the analysis of ideology can be traced at least back to Althusser. This tendency has subsequently flourished within the feminist wing, and in many cases has ended up leaving Marxist structuralism behind. Michele Barrett (1980), critical of the emergence of poststructuralist discourse analysis, tries to yoke ideological analysis more tightly to a material base in considering the construction of 'gendered subjectivity'.

Marxist feminist scholars face a difficulty in explaining the stubborn persistence of gender hierarchies under the emergence of capitalism, finding that they do not fit in with what is understood to be the *logic* of capitalism. But as Barrett says, 'an oppression of women that is not in any essentialist sense pre-given by the logic of capitalist development has become necessary for the ongoing reproduction of the mode of production in its present form' (Barrett, 1980: 249). The analysis of an ideology of gender, then, seems to fill the perceived gap. Feminist structuralists aim to locate this ideology within the necessity for reproducing the system. Thus, this kind of analysis of the perpetuation of ideology becomes another strain among the various meanings of social reproduction, structurally defined.

The term social reproduction as reproduction of gender ideology, specifically, has migrated out of the Marxian discourse into the mainstream of feminist analyses of women's subordination. For if the (reproduced) ideology of gender helps reproduce the system, it does so through its maintenance of that part of the social structure which relates economic existence directly to sexuality— namely, the sexual division of labour. In most current feminist thinking, this is the crux of gender relations as such.

C. *Social reproduction and sexual control /the sexual division of labour*

Feminist analysis points to the need to locate women's position, whether or not individual women are or ever will be mothers, within the hierarchy defined around the sphere of reproduction in the biological sense. Here, a general collection of observations is very common: (a) That women perpetuate the labour force through biological reproduction.[2] (b) That women are expected, via ideology, to provide the majority of the care of children, and thus of families, and thus are segregated into a less advantaged labour market. (c) That due to these constraints there is a sexual division of labour marked by the subordination of women. (d) That in order to maintain this subordination, societies exercise control over the sexuality of women. The term patriarchy enters into the discussion in relation to observations (b), (c), and (d). Patriarchy is continually reproduced via the sexual division of labour which is imposed upon the sexual division of human biology.

Feminists generally are concerned with these observations and sets of relations. Some Marxists and Marxist feminists then also tuck them back into the structural notion of social reproduction. Subordination based on sex, stemming from the possession by women of female sexual characteristics which include wombs, is seen to be integral to the reproduction of the mode of production as a whole. The reproduction of this order of things constitutes social reproduction. This links patriarchy directly with the mode of production, *even if* there is no essential reason why a given mode must, by 'logic', be patriarchal. The historical interaction of capitalism with patriarchy has been discussed in both First and Third World contexts (Hartmann, 1979; Brenner and Ramas, 1984; Beneria and Sen, 1981). It carries special salience in the much more overtly and formally patriarchal societies of the Third World.

If it can be said that the logic of capitalism is to draw more and more labour into the market, then under it patriarchy is illogical. However, Claude Meillasoux (1981) has provided a kind of scenario in which this illogic works itself out in the reproductive sphere. Elsewhere I have referred to his analysis of reproduction as identifying an 'interpenetrative mode of production' in which to explain the dynamics of reproduction, or even the laws of population (Clark 1987). In this mode, characterizing the current world order, an interpenetration of capitalist and pre-capitalist modes exists. Edholm *et al.* nicely summarize his thoughts (even while criticizing them): '[T]he logical implication of his analysis is that when capitalism has run out of other modes of production upon which it can practice the primitive accumulation of labour power, it will collapse since it cannot cover the cost of its own reproduction' (Edholm *et al.*, 1977: 108). Thus, we are back to social structural reproduction. The capitalist world system maintains itself by feeding on peripheral systems, until its internal contradictions bring it crashing down, according to the time-honoured Marxist formula.

But the question arises, how firmly is sexual control linked to the underlying productive mode? Is a patriarchal gender system essential to a capitalist or emerging capitalist mode of production? True, the preservation of such a system of sexual control defines the terms under which sexuality and biological reproduction maintain the labour force, an essential factor of production; and it maintains the grip of patriarchal social relations over the productive process. It interacts with the process of class formation in crucial ways as well. A patriarchal ideology of gender bolsters the maintenance of these relations. But the ideology identified here is one which has served for centuries to uphold a gender system far older than any one mode of production, any one kind of social structure. Does the social reproduction of a patriarchal society serve mainly to reproduce patriarchy? In what sense is a patriarchal gender system essential to the *continual* reproduction of the mode of production?

It is probably inappropriate to label the reproduction of patriarchy 'social reproduction'. Patriarchy is part of the relations of production and reproduction, but it is not the mode itself. We can imagine a world in which capitalism exists without patriarchy, just as we can historically identify worlds in which patriarchy existed without capitalism. Although not universally believed to have dominated the earliest human relations, patriarchy is also not coterminous with any one mode of production. A feminist effort to link biological and social reproduction needs to deal with this point.

A fresh approach to the incorporation of biological with social reproduction is found in recent work that suggests that the supposedly basic distinction between production and reproduction has been overdrawn. This dualism has been traditional in Marxist discussions, partly due to a lack of consciousness of the real value of women's work both outside and within the reproductive context, and partly due to an overdrawn conceptual dichotomy between 'public' and 'private'.[3] Recent thinking extends the concept of *work* to include more of what women do, beyond both the unpaid labour done in the family enterprise, and the unpaid labour done in the home, to the work women do in biological reproduction itself (Jeffery, Jeffery and Lyon, 1989). This thinking extends the valorization of women's contributions beyond that achieved in the discussions of another (but closely related) sphere of social reproduction: housework and family maintenance (which I shall examine shortly).

At this point, it becomes evident that a new space has recently been opened in the discussion of social reproduction as it relates to sex. Instead of just seeing women as the vessel of human labour power (defined by sex), we also see their biological labour (masked by gender assumptions) as work in itself. And this opens the discussion to a much larger consideration of the concerns appropriate to demography, as a study of biological reproduction: age and duration of marriage, pregnancy, foetal wastage, childbirth, lactation, maternal morbidity and mortality, infant and child mortality, birth intervals, maternal depletion, sterility. These concerns can be seen as part of the working conditions of ordinary women, applying not only to their reproductive labour per se, but also to most of the work — both 'public' and 'private' — that they do during their reproductive careers. In the second half of this paper, I discuss a few of these demographic concerns in relation to India.

All of these demographic concerns and topics of study, within their various socioeconomic ramifications and contexts, can theoretically be linked with the more traditional Marxist concerns with at least one degree less of separation than before when we consider them as aspects or direct characteristics of socially necessary work. Gender is then seen as more than the ideology that bolsters a patriarchal form of gender relations and normatively dictates rigid roles to be played by the sexes. Its material basis is revalued. Thus, even as it is being transformed the sexual division of labour can be re-evaluated.

Gender is today in much of the world a very unequal power relation, based on sexual control. But if the biological work that can only be done by one sex were more thoroughly valorized conceptually, gender could be envisioned as no longer being an ideological or even a structural instrument of oppression, deriving from patriarchy. A valorized concept of biological work, translated into real value and power, could lead gender to become a culturally malleable expression of an essential difference,[4] a difference that need not carry the acute disadvantages it does in the present.

Gender is historically and currently, in many places and certainly throughout much of India, a harsh yoke, based very firmly on sexuality and its control, biological reproduction and its control, and the appropriation of the work that women do. So many categories of women's contributions are devalued that they have been almost invisible. Feminists have tried to shed light on these in various categories, and one result has been the identification of homemaking activities with social reproduction.

D. Social reproduction and housework of all kinds

Labelling family maintenance activities as social reproduction has become perhaps the most popular, most widespread use of the term today. Housework and childcare issues have their own politics, which must be worked out in the personal arena of many feminists' own lives. Housework was one of current feminism's earliest theoretical concerns.

Correspondingly, a body of literature by not only feminist scholars, but also by mainstream economists and sociologists, has developed on domestic labour and household production, as ramified by surrounding economic conditions. Marxist feminist scholarship points to the stage of incorporation into capitalism as determinative of the nature and extent of domestic work. Processes of class formation correspondingly have crucial effects on the extent of domestication of various kinds of labour. In an agrarian economy, household production merges with the work women contribute to the family enterprise. Women perform childcare and home maintenance activities alongside home production of items that would be market commodities in more advanced sectors. The collection of data on these activities is an important contribution to development literature (Dixon, 1978).

Even in advanced economies and sectors, women do most of the maintenance activities and childcare when they are at home, even if this is after a day's work outside the home. The burden of the double day even in the most advanced sectors of the world has elicited serious recent scholarship (Hochschild, 1989; Hartmann, 1987).

By themselves, however, these activities do not constitute social reproduction in a structural sense. The term social reproduction as applied to them is used in a somewhat colloquial way. Instead, they are part of the broader pattern of the sexual division of labour extrapolated from biological function and gender ideology, which we discussed earlier, further ramified by class formation processes.

But the home forms the essential nexus for the reproduction of ideology—not only gender ideology, but also the ideology supporting class relations and the ownership and transmission of private property. (Here the work of Pierre Bourdieu needs to be consulted). Children are socialized in the home; in the patriarchal family, rewards and punishments are meted out to adult members as well; in household production, ideology also maintains a subordinate labour force. The intensity of identity formation and maintenance, based on the social relations located in the household-family context, cannot be overlooked in an analysis of various forms of social reproduction.[5] In fact, the term social reproduction has sometimes been applied directly to socialization in the home.

This aspect of home-based social reproduction is well described by Maria Mies in her work on the Indian lacemakers of Narsapur (Mies, 1982). Class as well as gender is maintained through an ideology of seclusion, which separates the lacemakers from each other and from potential solidarity or class or gender consciousness. What is reproduced is both an ideology, or an acute consciousness of certain rules that cannot be questioned, and at the same time a particular *lack* of consciousness which could threaten the maintenance of this class of producers at the level of exploitation they suffer. Here we observe forces that help reproduce the social structure by reproducing certain relations of production.

There is another element of importance within the focus on the home, having to do with the reproduction of an essential element within the capitalist mode of production, namely the accumulation of capital within families via inheritance. Engels believed this process was essentially linked to gender, and has enjoyed a rediscovery by feminist scholars as a result.

E. *Social reproduction and the transmission of private property*

Engels' essay on 'The Origins of the Family, Private Property and the State' (Engels, 1975) tied all these institutions' origins together. The subordination of women, he argued, originated when ownership of the means of production became private, because private property was controlled by men. It was the need to pass on male property to the next generation that necessitated control over women and their sexuality, so that paternity could be verified. Private property was the yoke that bound women, as well as being the source of capital that could be transmitted and accumulated. The process of class formation and of the succession of modes of production followed; private property was thus the source of the capital that made the development of capitalism possible. The elimination of private property under socialism, then, should free women.

Experience with would-be socialist governments has called the last assumption into question. But all the more, the Engels thesis left a partial vacuum by not explaining adequately why property and the family were of necessity patriarchal. Scholars within a number of disciplines, including archaeology, have responded to the challenge of doing so. The question is an historical one.

Engels called crucial attention, however, to the long-term dynamism of family-property relations: the simultaneous reproduction of these relations along with life itself, that is, their generational mobility. With private property, inheritance, and accumulation of capital within the propertied class, we have elements of social reproduction, maintained and carried forward by ideological conventions, that are structurally essential to the developing mode of production.

Systems of gender relations have played an important role in maintaining and pushing forward the relations of production used by successive modes. Gender and political economy have often supported each other in a powerful way; it is important that the nature of the relationship is historical. One of the quandaries in dealing with the concept of social reproduction, as the reproduction of society, is a temptation towards functionalism that seems implied in the very concept. The historical perspective must cure this tendency.

An important historical theme in the reproduction and transformation of society is that of population growth. This represents biological and social reproduction in history. It also relates closely to other historical themes in social reproduction, such as the accumulation of capital, and the formation and articulation of class relations. Recognizing the historical nature of social reproduction leads us to consider the historical experiences of specific societies.

II. GENDER AND GENERATIONAL SUCCESSION: AN APPROACH TO INDIAN HISTORY

In 1981, Lourdes Beneria and Gita Sen published an article in *Signs*, bearing the provocative title 'Accumulation, Reproduction, and Women's Role in Economic Development: Boserup Revisited'. The title ties together several important historical themes.

The article offers a re-evaluation of Ester Boserup's contributions to the women and development field. Embedded in the article almost in passing, however, are rudiments and suggestions of a theory of population history, with special reference to women's lives: a theory which considers both gender and class, development and underdevelopment under colonialism. This theory is projected onto the empirical canvas of India. In the rest of this paper, I will attempt to carry the development of such an historical theory of population and development a few steps further, also in relation to India.

A theory of population and development history with reference to gender is very much needed, but as yet has been only sparsely developed. Such a theory could help us to articulate better the full extent of the social impact of Western hegemony on Third World countries. It could suggest explanations of the whys and wherefores of world population growth. And it could help illuminate a little more of the forgotten history of the world's women.

Although gender roles play a part in the design of much contemporary demographic research, the theoretical basis for such research is often lacking in historicity. Nor does such research often delve deeply into the way class dimensions of the most basic population processes—fertility and mortality—relate to their gendered dimension. Without any consideration of class or other elements of political economy, such as the stage of integration of the society into a capitalist world economy, many elements of the gendered dimensions of demography are inadequately understood.

The Beneria-Sen article, oriented around a critique of Boserup, only briefly deals with population processes. We learn what Boserup failed to do in her work on women and what needs to be done better. But in grappling, even briefly, with issues of Third World population growth and density under the introduction of capitalist relations, the article's implications are potentially very important for the study of reproduction and its history.

What are these implications, and why are they important? The writers fault Boserup's critique of Malthus (for which she is famous in demographic circles) for its lack of a class analysis or a mode of production framework. They point to the need to provide, as Boserup does not, a 'coherent analysis of the interconnections between the social process[es] of accumulation, class formation, and changes in gender relations' (287).

These three processes are both historical, and relevant to the forms of social reproduction we have examined. The authors claim that there is an essential link between the development of the economic and social structures and relations of capitalism, on the one hand, and the position of women, on the other. How do these links work themselves out historically?

I will argue that this occurs directly through the nexus of biological reproduction, and that social structural and biological reproduction are so closely tied that they require joint discussion.

Let us sketch the theory which is suggested in the article. There are two sets of passages which contain suggestions of how we ought to examine the history of India in light of interconnections between population growth, changes in modes of production (that is, accumulation and class-formation processes), and gender relations.

The first refers specifically to the colonial period, and to population processes which played a part in the development of relations of production characteristic of imperialism. The authors criticize Boserup when she fails to 'confront the causes of growing population density . . . [This] involves ignoring effects . . . of the alienation of land and its private appropriation during the colonial period... on population growth and density' (285). 'The indirect effects have been felt in most regions where the privatization of land, labour, and subsistence have generated incentives for higher fertility among peasants' (286).[6] There is a need to view correctly the causes of 'women's worsening situation under conditions of rapid land alienation and class differentiation' (287).

These remarks raise challenging historical questions. What was the extent of land privatization under colonialism, and how did changes in agrarian relations connect with population growth and density? Specifically, what is the link between land privatization and increased fertility? What were the important periods within this relationship, and how did these events affect different parts of India differently as colonial penetration advanced? What have been the differential effects on women's situation by class?

Research I have done on Gujarat has addressed some of these questions (Clark, 1983). In Gujarat's history, the formation of class and gender hierarchies, and resultant population trends, are more explicable together than separately. Land privatization in the hands of an emerging elite is linked to proletarianization and high population growth among lower-class peasants. To understand more of India's population history, many more regions need to be examined in detail.

The second set of remarks which is important for piecing together the implicit theory focuses on the relationship between class and gender. Beneria and Sen find this relationship to be a contradictory one. 'The poor peasant household may survive off [sic] the continuous pregnancy and ill health of the mother, which are exacerbated by high infant mortality. The mother's class interests and her responsibilities as a woman come into severe conflict' (296). With the spread of enlightened family-planning programmes,

[t]he reduction in infant mortality, improvement in health and sanitation, and better midwife and paramedic facilities can give poor rural women more options than having to resolve class contradictions through their own bodies. Such programs, however, clearly cannot be a

panacea for the basic problems of extreme poverty and inequality in land holding; the contradictions of class and capital accumulation in the countryside can be resolved only through systemic social change (297).

Are class and gender really in conflict? Even in Africa, where fertility levels are among the highest in the world, women are not 'continuously' pregnant. However, Jeffery and Jeffery (in this volume) show that, combining pregnancy and lactation, women's biological work is almost continuous in their North Indian rural research area. What kind of conflict between women's different roles can this imply?

I believe what Beneria and Sen most usefully can be interpreted to mean here is that among the very poor *fertility is in conflict with mortality*. If a woman's 'class interests' can be interpreted to equal fertility—the sheer production of children to ensure the continuance of the group; and if her 'responsibilities as a woman' refer strictly to the preservation of her children's lives, and even her own life—in other words, survival—this makes complete sense demographically. Mortality is, of course, the limit to life. The survival of children and the welfare of mothers are threatened in areas of the world fraught with high mortality. If some classes or regions face a greater threat than others, there women's work of biological reproduction is made far more difficult.

But of course classes are not reproduced only by the birth of babies to women identified with those classes—any more than the labour force relies on fertility alone to supply it with new workers at any point in time. The history of class formation is distinct though overlapping with strictly demographic processes. There are also questions about the class location of women as distinct from their husbands. And there is some tendency here to identify women as vessels of labour, a problem referred to earlier in the discussion of how biological reproduction has been identified with social reproduction.

The notion of resolving class contradictions through women's bodies is extremely suggestive, however. It suggests the possibility of exploiting a dialectical framework to explore more aspects of the relations within which women are embedded. Exploring the relations between mortality and fertility in this way is very natural to demography. It is also problematic, and exploring this problematic and attempting to clarify the issues within it require some attention to basic concepts in demography, which I shall do in the next section.

The partial theory of population and development history, only sketchily suggested in Beneria and Sen's article, can now be more clearly stated. It relates to the original observation of Marx that 'every mode of production has its own laws of population' (O'Laughlin, 1977). The penetration of colonialism into peasant economies altered relations of production, in the process of beginning a shift in the overall mode of production. Land alienation and its accumulation in private hands went hand in hand with intensified processes of population growth and impoverishment for lower classes. In situations of extreme impoverishment, the biological task of reproduction, carried out by women, has at some times and places become intensely burdensome to them. It is a burden they have often borne at the cost of their lives. They experience firsthand the thesis and antithesis of vital processes. Yet with all the threat to poor women's and children's lives, under colonial and post-colonial relations of production there are pressures that drive population growth onward.

There is more to say about this theory than can be discussed adequately here. It is a partial historical theory, needing further elaboration. One point that must be made here is this: the mode of production into which colonialism historically has moved Third World countries cannot be well understood by looking at those countries in isolation. They have been moved into an interpenetrative mode of production on a world scale, as part of their integration into a global world economy under western hegemony.

As this process changes the relations of production at the societal level, the life and death struggle that women endure they experience at the level of the family and the community. They are concerned with the relations between and continuation of the generations. The conceptual privatization of these experiences and concerns has been responsible for our overlooking much of their importance for history.

In what follows I choose to interpret the demographic dialectic that first needs to be dealt with as the one between fertility and mortality that takes place at a societal level. Because class can never be seen as a closed category, it is far more difficult to analyse demographically. But since class dynamics within a population should be understood as crucial to the driving forces of its growth, the demographic issues relating to class call out to be carefully addressed. The question of class becomes relevant at a level of disaggregation just below the societal level. But gender is relevant at both the societal level and the level of class formation and maintenance.

III. A GENDERED POPULATION HISTORY OF INDIAN REGIONS

Population growth has inherently gendered components within it. The processes which combine to produce population growth (or the lack of it) are fertility, which self-evidently involves gender, and mortality. Feminist scholars in general have become keenly aware of gender bias in the treatment of females, which may lead to excess female deaths as it does in India (Miller, 1981). What is not always understood, however, is that there is a gendered dimension of population *growth* which stems as inescapably from mortality as it does from fertility. It derives from the inverse of female mortality: the extent of female survival to adulthood, with all of its social as well as physical determinants.

Population growth is thus produced or inhibited by the combination of biology and gendered social determinants. In other words, population growth is partly historical and social; and it is so, in large share, by way of gender. To say this is not to equate gender and sex, or to say that 'biology is destiny'. Gender systems, operating at the level of the family and community, determine much about the impact of biology, meting out considerable care for women in some communities and classes, intense pressure in others. Even at the most aggregated level of a whole society, however, something of the character of women's life experiences can be read off from the bare demographic record. The demographic concept which most clearly encapsulates the gendered dimension of population growth is called net reproduction, as I will explicate below.

Following up on my earlier studies on India's regional historical demography, I examine here several different demographic regimes which have prevailed in different regions of India, rooted in their particular political economies. The three regions examined here are the old Bombay Presidency, Madras Presidency, and United Provinces—today's Maharashtra and Gujarat, Tamil Nadu and Andhra Pradesh, and Uttar Pradesh, respectively.

The demography of these three regions over the fifty-year period between 1881 and 1931 has been explored in research reported on in two earlier articles (Clark 1987, 1989).[7] These decades were a particularly harsh period in India's demographic history. Mortality was very high, although it varied considerably between regions. The worst mortality over the half century was found in the United Provinces, today's Uttar Pradesh. There the population barely grew at all over fifty years. The next worst mortality was found in Bombay, and that of Madras was markedly better. Female mortality followed different patterns than male mortality, although the way these sex-specific patterns differed varied between regions.

In Madras, excess female mortality occurred by and large only during the childbearing ages. It was not evident in infancy and early childhood. In fact, in infancy male mortality well exceeded female, as it does in populations where natural factors predominate and females are not exposed to excess neglect. Madras mortality has persistently been more like that of western countries than those of the other two regions. My best estimate of the sex ratio of persons surviving to age 30 in the middle of this period (Clark,

1989, p. 145) is about 1032 males to every 1000 females (or as expressed in India the other way around, 969 females to every 1000 males). Since this is below the sex ratio at birth of 1055, male over female, it does not suggest early excess female mortality.

In Bombay there was a high degree of excess female mortality in the earliest ages throughout the half century. The sex ratio of survivors to age 30 in the middle of the period I have estimated to have been between 1160 and 1200 male over female, or between 833 and 862 female over male.[8] This range of ratios was a product of pronounced sex discrimination in early childhood and its continued ill effects on female survival during early childbearing. These patterns, associated with the strongly patriarchal North Indian cultural milieu, were more clearly evident in Bombay than in U.P. in this period because Bombay's overall, systemic mortality was not so high.

In U.P., sex-specific patterns shifted from favouring females to survive during three decades of crisis mortality, to favouring males as conditions finally began to improve. Infant mortality was worse among males than females throughout the crisis period, as it is in 'ordinary' western cases. What was not ordinary, however, was the truly devastating extent of mortality in general. Excess female infant and child mortality only showed up in U.P. when the level of overall mortality improved at the end of the period. (For the levels of mortality by decade, by province, see the life tables in Clark, 1989.) At mid-period, the sex ratio of survivors to age 30 was about 1006 male over female, or 994 female over male.[9] This was far lower (male over female) than the average sex ratio at birth, thus showing a heavy toll of mortality on males. Survival of both sexes to adulthood was at its lowest in U.P.

Let us now relate survival more specifically to the process of population growth. However many women remain, relative to men, at close to the middle of the childbearing span, they are the ones who bear the burden for continuing the population: who must do the work of reproducing the society in a biological sense.

Table 1 represents an analysis of the burden that they bore during this period, differentiated by region.

The Table is about population growth and its essential components: fertility and female mortality (shown as its inverse, female survival). In trying to find the possible ranges within which each of these components may have fallen, a number of conventions and assumptions have been used which are briefly explained here.

The anchor for this table is the average annual growth rate of each region's population over the fifty-year period. Bombay's growth rate over the period was 0.69 per cent per annum; U.P.'s was 0.24 per cent; and Madras's was 1.03 per cent.[10] These growth rates have been transformed into Net Reproduction Rates (NRR) for each region, using the formula NRR = e^{rt}, where r is the annual rate of growth and T is the mean age at childbearing. (The assumption was made, generalizing from a considerable range of Indian data, that the mean age at childbearing across India averaged twenty-seven years.) The equation NRR = e^{rt}, solved for each region using its growth rate, computes generational growth. The NRR is the amount a population would grow in a generation (twenty-seven years long) if the growth rate remained stable.

The Net Reproduction Rate is also a gendered measure. It is equivalent to the number of daughters borne by the average woman, daughters who then manage to survive to the mean age at childbearing to begin producing the next generation. In itself it is worth close examination, for one's expectation is that Indian fertility was historically fairly high; but the number of surviving daughters per woman, as shown in the Table, was extremely low. NRR is a fertility measure that takes account of female mortality, showing its toll on women's ability to carry out their reproductive work.

(There is an irony that becomes clear in thinking through the relationship of net reproduction to the usual concept of women as vessels of labour. As bearers of daughters, women are, of course, vessels of female labour—in some cases highly disadvantaged, yet still essential to the continuance of reproduction in any sense. To disadvantage essential producers is at the heart of both sexual control and class dominance.)

The Net Reproduction Rate as a demographic measure can be cleared of the effects of female mortality by being divided by an estimate of female survival to the mean age at childbearing.[11] It then yields the Gross Reproduction Rate (GRR), the number of daughters (surviving or not) which the average woman would produce over her childbearing career if she survived to complete it. Inflating this number by one plus the standard sex ratio at birth (1.055 males per female) yields the Total Fertility Rate (TFR), which is the number of children of both sexes the average woman would bear under the same condition. These are fertility measures without direct reference to mortality. They begin to represent the high fertility we expect to find in India. We are working backward, from reproduction as modified by mortality, to the original production of babies.

Some women did not bear as many children as others, though they may have survived to the end of their childbearing span, because their husbands did not. The assumption made here is (based on age-specific widowhood data in the Census) that of those women who lived out the childbearing careers about 17 per cent were widowed before completing their potential childbearing careers. And remarriage was extremely rare. One might express it that only about 83 per cent of married women's total childbearing capacity was used as a result. The TFR is inflated by the reciprocal of .83 in order to provide an estimate of the Total Marital Fertility Rate (TMFR).

This final rate represents the total number of children borne by the average married woman who survived to the end of her childbearing span unwidowed.[12] The TMFR is a fertility measure that shows what share of all childbearing had to be borne by actual survivors of the threat of death to themselves and their husbands to sustain population growth at observed rates.

The average annual growth rates over the fifty-year period, as transformed into NRRs, anchor the table in the following way. The NRRs are held constant across all three panels, and alternative pairs of fertility levels and female mortality levels (based on different outcomes of the same life tables from Clark, 1989) are then fitted into them. The alternative pairs express the possible ranges of fertility and mortality that could have made up these NRRs, and hence have composed the observed rates of population growth. The third panel represents the averages within each range, and I now discuss these average estimates.

Clearly, the southern part of India (Madras) had both the highest growth rate, and the lowest fertility and female mortality (because the highest female survival). Bombay had lower growth, intermediate mortality, but the highest fertility. U.P.'s female mortality was the highest of all; its growth was the lowest, and its fertility intermediate. With its extremely low female survival rate, U.P.'s fertility rate (carried forward by the survivors) was still very high.

Considering the findings more specifically, we see that in U.P. only about one-third of all the females ever born survived to age 27. Yet the average number of children borne by all women was 6.55, and by women whose marriages were not broken by widowhood, 7.90. This burden of reproduction was borne by women in conditions of the highest overall mortality, which affected both males and females (the sex ratio at age 30 was less masculine than the sex ratio at birth). One can well imagine that in such a high mortality regime, the health of many of the surviving mothers was compromised, so that their ability to bear children was weakened. High fertility, in turn, undoubtedly led to high mortality of infants and mothers. In such an environment, the fertility level was high indeed.

In Bombay, the survival rate of women tells a somewhat different story. Only 36 per cent of women survived to age 27, but men did better. Since the adult sex ratio was estimated to be highly masculine (at around 1160, male over female), more than 39 per cent of all males ever born managed to live to grow up. This was not a good mortality regime, but was somewhat better than U.P.'s, where only a third of the population survived. The higher female than male mortality related strongly to sex discrimination practices.[13] Yet surviving females bore the highest number of children of all regions, and marital fertility was above eight children per woman.

In Madras, women's fertility was almost as high as in U.P. But it was so under better survival conditions, and with an adult sex ratio not much different from that found at birth. Neither did the highest mortality threaten Madras women, nor did they suffer discriminatory behaviour to the extent it was found in the north and west. A region where 42 per cent of all those born survive to age 27 cannot be called a low-mortality regime. But it was less harsh than the others.

Why was this so? Over against the almost purely cultural explanation for regional differences in demographic regimes, proposed by Dyson and Moore (1983), I propose a more historical one. Under the advancing colonial penetration of India, I believe that the south may not have become either so poor, so peripheral, or so proletarianized as the north or west. These are three alternative, but potentially mutually reinforcing, hypotheses.

In India, increasing integration into the world economy under western domination was a powerful historical force. How did it differentially affect the separate regional economies? Part of the answer ineluctably appears in the historical demography of those regions. Fuller explanations remain to be worked out in future research; but the lines of inquiry that suggest themselves would surely be oriented, in part, toward the different land-tenure systems that were evolved for the different regions for administrative convenience. It is generally known that these tenurial arrangements had particularly harsh effects on the political economy of the north, and were more egalitarian in the south. The different ways the regions were integrated into wider markets would then be important to examine. These approaches form a background to the history of class formation particular to each region, with its demographic implications.

In my earlier discussion of these regional demographic regimes, I note that epidemiological differences were more outstanding in determining demographic rates than were cultural ones (see Clark, 1989). This is not to deny that cultural differences played a role, but simply to say that they do not explain everything. Epidemiology itself, moreover, is no more useful as a unique and insular explanation than culture. Historical forces dampen or intensify regional tendencies, and may sharply influence the prevailing disease environment.

Cultural differences between the regions affected the gender systems characterizing them, and these clearly played a role in their reproductive histories. Subsets of southern Indian society were characterized by less harsh, less patriarchal gender systems than much of the north; and this is reflected by the much closer mortality rates between males and females in Madras than elsewhere.

But this is not particularly remarkable or new. Several rather more remarkable observations emerge from a comparison of these regions' demographic records. In Madras's demographic record, a gentler gender system marched in step with a gentler epidemiological situation, so that women's overall reproductive burden was lighter and less fraught with threat than elsewhere. In sharp contrast, a potentially very harsh gender system in U.P., whose excess female mortality effects have been strongly felt in recent decades, was modified in its demographic effect during this historical period by an even harsher epidemiological situation. Somewhere in between, Bombay exhibited the clear demographic effects of a harsh gender system within an epidemiological situation poised midway between those of south and north.

Of course, the regions did not each have uniform gender systems within themselves. Gender systems varied by caste and class, and had varying demographic effects across social levels.[14] Within the ongoing process of class formation, differential demographic patterns articulated self-reinforcing class differences. Thus, the interconnections between class and gender have been both complex and dynamic, and have been inextricably linked to the twin processes of development and underdevelopment.

CONCLUSION

We cannot really understand social formations, or the reproduction of societies, outside an historical context — a context that displays them as moving and changing, reproducing themselves in ways that speak but are never quite the same.

The concept of social reproduction, as the reproduction of society based on its underlying mode of production, needs to be lifted out of the thickets of abstraction. A better understanding of historical forces such as population growth, the relations of production and reproduction within which such growth occurs, and the unfolding of concomitant social dynamics, can help us demythologize a would-be historical discourse grown heavy with unexamined assumptions. I have tried in this paper to move between the theorizing about social reproduction, on the one hand, and social and reproductive history with its regional particularities, on the other, at the same time emphasizing the theme of material continuity.

The over-distinction between public and private spheres of life has often masked essential features of the production-and-reproduction totality. A mode of production is overtly characterized by its technology—its productive forces. But history suggests that the management of life and society are equally crucial for us to understand—the relations of production and reproduction. Both elements coexist inextricably, and are equally 'structural' with respect to the basis and continuance of society.

This connection is analagous to, and reflected in, the connection within women's lives (within their very bodies) between public and private spheres, production and reproduction. Gender cannot be seen as a relation of production separate from, and somehow lesser than, other relations of production such as class.

Just as women are the reproductive class, so is gender the reproductive relation. As such, it engenders and is necessary to all the reproductive relations a society must sustain to continue and renew itself. Gender is at the matrix of the reproduction of both life and society. Therefore historical studies of women's life conditions are potentially far more relevant to the understanding of the history of society than has been generally accepted up to now.

In this discussion of Indian historical demography as an exemplar of social reproduction, suggestive of much further research that needs to be done, I have tried to breathe some of the life back into historical processes of living and dying as they occurred in the major regions of India. By imagining other lives out of their emergence from the record through whatever evidence is available, and then by trying to understand what their passage has meant in a larger context, we can enlarge our sense of what history is.

NOTES

1. A large share of this privatization of women is conceptual. Women of the lower castes and classes have worked outside the private sphere for centuries. Their numbers and proportions have increased over time. But the dominant concept of the sphere in which women's lives are lived has traditionally been the private one. This has been appropriate for a large share of elite women and reveals the elite bias of Indian history in general.
2. Edholm *et al.* point out that the maintainance and expansion of the labour market at a point in time is not dependent only on childbearing patterns, but also on migration and shifts in roles. In the broadest sense, and in regard to the succession of the generations, however, biological reproduction is basic.
3. Some of it is due to the persistent dualism of western thought, certainly any school of it deriving some of its mental equipment from Hegel. Yet a dialectical framework has not yet been as rigorously applied to the dualism between 'public' and 'private' as it might be, so that there is potential for exploiting it further.
4. It should be clear that I am referring directly to reproductive dif-

ference, and not to cultural or psychological differences which have engaged some liberal feminist writers. Such differences may exist, but they articulate socially variant responses to the underlying difference, which does not require them.

5. Much thinking on this topic has been done on advanced capitalism. Christopher Lasch's *Haven in a Heartless World* provided an interesting and controversial example.
6. The authors reference Mamdani's *The Myth of Population Control*. This has now become outdated by the onset of another phase of demographic change in the Panjab region he studied (Nag and Kak, 1984). But the book did document a particular stage in the historical relationship between the advance of capitalist relations and population growth in that region.
7. I do not present technical details here; the methods used and the evaluation of data are discussed in detail in the 1989 article. The Appendix to the table presented here describes procedures used for producing the basic estimates for this paper.
8. Calculations made for Clark, 1987, but not shown there; can be provided on request.
9. Based on the preferred pair of life tables for the 1901–11 decade.
10. I attempted adjusting for migration and found that it made a difference only at the fifth decimal point level, so I considered it negligible.
11. Estimates derived from the life tables computed for Clark, 1989, as described in the Appendix.
12. There is no assumption made about the additional share of total childbearing relatively healthy women bore on account of some women's early sterility, though such an assumption might usefully be made.
13. These have been examined in detail in Clark, 1983.
14. As seen for central Gujarat in Clark, 1983. The lower-class castes had higher fertility and faster reproduction than the propertied castes; the landed caste curbed its own reproduction by practices that caused excess female infant and child mortality directly or indirectly.

TABLE 1
FERTILITY AND FEMALE MORTALITY AS COMPONENTS OF POPULATION GROWTH IN THREE INDIAN REGIONS, 1881–1931

A. Finding fertility by using survival values from female life tables:						
	r	NRR	$l_{(27)}$	GRR	TFR	TMFR
Bombay	.0069	1.20	0.348	3.45	7.09	8.54
U.P.	.0024	1.07	0.324	3.30	6.79	8.18
Madras	.0103	1.32	0.407	3.24	6.66	8.03
B. Finding survival by using fertility estimates from female life tables:						
Bombay	.0069	1.20	0.378	3.18	6.53	7.87
U.P.	.0024	1.07	0.347	3.08	6.33	7.63
Madras	.0103	1.32	0.441	2.99	6.14	7.40
C. Averaging within the ranges of fertility and mortality:						
Bombay	.0069	1.20	0.363	3.31	6.80	8.20
U.P.	.0024	1.07	0.336	3.19	6.55	7.90
Madras	.0103	1.32	0.424	3.12	6.41	7.72

Measures:

r = average annual rate of population growth
NRR = Net Reproduction Rate, derived from population growth rate
$l_{(27)}$ = Proportion of females born surviving to age 27
GRR = Gross Reproduction Rate: NRR / $l_{(27)}$
TFR = Total Fertility Rate: GRR x 2.055
TMFR = Total Marital Fertility Rate: TFR / .83

Source: Census of India. See Appendix for explanation of life table construction methods. See text for explanation of assumptions and measures.

APPENDIX

The data used are from the six decennial Censuses of India taken from 1881 through 1931. Population totals and age distributions for the three regions are taken from Mukherjee, 1976. Migration data by region were collected from the original census volumes. Alternate life tables were generated from these data for each decade, using different assumptions about the age and sex distribution of mortality; methods are spelled out in Clark, 1989. Fertility levels were derived from each life table by regression.

The life table values from that work which are used here are from the life tables designated in that paper as preferred ones for each decade. These are as follows: for Bombay, the South Asia tables for 1881–91, 1901–11, and 1921–31, and the Crisis tables for 1891–1901 and 1911–21; for U.P., the South Asia tables for 1881–91 and 1921–31, and the Crisis tables for 1891–1901, 1901–11, and 1911–21; and for Madras, the Model South tables for all decades except 1911–21, for which the Crisis table was used.

These sets of life tables are used here as follows. The mortality measure used is female survival to age 27 (assumed to be the mean age at childbearing), written as $l_{(27)}$ in life table terminology. For Panel A, values of this survival rate from each of the preferred tables were summed and a fifty-year average was derived for each region. A GRR was then derived from this measure by dividing NRR by this $l_{(27)}$. The fertility measure used is the Gross Reproduction Rate, GRR. To provide estimates of this rate for Panel B, the values of GRR from each preferred life table were likewise averaged over fifty years for each region. An $l_{(27)}$ was then derived from each GRR by dividing NRR by it. Measures for Panel C are averages within the ranges of fertility and mortality thus derived, checked for internal consistency.

BIBLIOGRAPHY

Michele Barrett, *Women's Oppression Today*, London: New Left Books, 1980.

Lourdes Beneria and Gita Sen, 'Accumulation, Reproduction and Women's Role in Economic Development: Boserup Revisited', *Signs: Journal of Women in Culture and Society*, 7(2): 279–98, 1981.

Ester Boserup, *Women's Role in Economic Development*, London: George Allen & Unwin, 1970.

Johanna Brenner and Maria Ramas, 'Rethinking Women's Oppression', *New Left Review*, 144: 33–71, 1984.

Alice W. Clark, 'Mortality, Fertility, and the Status of Women in India, 1881–1931', in Tim Dyson, ed., *India's Historical Demography: Studies in Famine, Disease and Society*, London: Curzon Press, 1989.

——, 'Social Demography of Excess Female Mortality in India: New Directions', in *Economic and Political Weekly*, 22(17), Review of Women Studies, 25 April 1987.

——, 'Limitations on Female Life Chances in Rural Central Gujarat', *Indian Economic and Social History Review*, 20:1, January–March, 1983.

Ruth B. Dixon, *Rural Women at Work: Strategies for Development in South Asia*, Baltimore: Johns Hopkins University Press, 1978.

Tim Dyson and Mick Moore, 'On Kinship Structure, Female Autonomy and Demographic Behaviour in India', *Population and Development Review*, 9: 35–60, 1983.

Felicity Edholm, Olivia Harris and Kate Young, 'Conceptualising Women', *Critique of Anthropology*, 3(9–10): 101–30, 977.

Friedrich Engels, *The Origins of the Family, Private Property and the State*, New York: International Publishers (reprint edition). First edition published in 1884, rpt. 1975.

Heidi Hartmann, 'The Unhappy Marriage of Marxism and Feminism: Towards a More Progressive Union', *Capital and Class*, 8, 1979.

——, 'The Family as the Locus of Gender, Class, and Political Struggle', in Sandra Harding, ed., *Feminism and Methodology: Social Science Issues*, Bloomington: Indiana University Press, 1987.

Arlie Hochschild, *The Second Shift: Working Parents and the Revolution at Home*, New York: Viking, 1989.

Patricia Jeffery, Roger Jeffery and Andrew Lyon, *Labour Pains and Labour Power: Women and Childbearing in India*, London: Zed Books, 1989.

Christopher Lasch, *Haven in a Heartless World: The Family Beseiged*, New York: Basic Books, 1977.

Mahmood Mamdani, *The Myth of Population Control*, New York: Monthly Review Press, 1972.

Claude Meillasoux, *Maidens, Meal and Money*, Cambridge: Cambridge University Press, 1981.

Maria Mies, *The Lace Makers of Narsapur: Indian Housewives Produce for the World Market*, London: Zed Books, 1982.

Barbara D. Miller, *The Endangered Sex: Neglect of Female Children in Rural Northern India*, Ithaca: Cornell University Press, 1981.

Sudhansu Bhusan Mukherjee, *The Age Distribution of the Indian Population: A Reconstruction for the States and Territories, 1881–1961*. Honolulu: East-West Population Institute, 1976.

Moni Nag and Neeraj Kak, 'Demographic Transition in a Punjab Village', *Population and Development Review*, 10(4): 661–78, 1984.

Bridget O'Laughlin, 'Production and Reproduction: Meillasoux's *Femmes, Greniers et Capitaux*', *Critique of Anthropology*, 8, 1977.

5

Social Classes and Gender in India: the structure of differences in the condition of women

KALPANA BARDHAN

Some feminist studies in the Indian context view women as a more exploited *class*, in the sense of having to render more labour which is under (or un)remunerated, gaining disproportionately less than they contribute to the accumulation of capital, relative to men. The report of the Committee on the Status of Women in India partly takes this stand. (The opposite stand is illustrated by the mainstream trade-unionist view that focusing on gender divides the class struggle and defeats itself.) This view of women as a more exploited class sees this being accentuated in the economic growth process, in agriculture and other sectors, with the rising proletarianization of communities that used to be self-reliant within household economy and intra-community exchange relations. Other studies view patriarchal institutions in the domestic arena as being at the root of most forms in the oppression of women, irrespective of their class location. The magazine *Manushi* leans heavily towards this view.

Both these views leave unaddressed certain problems of both empirical fit and conceptual adequacy in the context. Concepts of domination, exclusion, segregation, powerlessness, etc. have to be qualified for non-polarized relations which have complementarity of roles and tasks, and at least some shared interests and joint goals even if the division of gains and losses is systematically unequal by sex. Behind-the-scene indirect leverages at least partially modify patriarchal domination and even the strictest segregation of domestic and social arenas; love and loyalty—parental, filial, conjugal—are counterforces submerging conflicts; and even the balance of power changes over the female life-cycle with accrual of returns to the earlier emotional investment in children, with the formation of what Margery Wolf describes as the 'uterine family' in defensive reaction to the harshness of patriarchy towards young in-marrying women. The second view—of patriarchy as the primary source of oppression—has not taken adequate account of these checks, balances, and counterforces within patriarchy, aside from implying that class exploitation is either sex-neutral or proportional to domination of a patriarchal or any other kind.

Basing the concept of class on that of domination poses problems for either view, as does the implicit assumption of uniformity in the hold of patriarchal institutions, even within a country. Erik Wright notes two conceptual problems. First, although exploitation generally involves some form of domination, it specifically involves the appropriation of the surplus generated by labour over own subsistence. If the direct beneficiary of the surplus generated by female labour is the capitalist enterprise (farm or firm) or the landlord/moneylender, not the males within her family or community or similar families, then it is best characterized as capitalist or feudal exploitation, even if patriarchal customs aid it, as indeed they do.

Exploitation intrinsically implies a set of opposing material interests. Second, domination-centred concepts of class tend to slide into what can be termed the 'multiple oppressions' approach ... Societies, in this view, are characterized by a plurality of oppressions each rooted in a different form of domination—sexual, racial, national, economic, etc.—none of which have any explanatory priority over any other. Class, then, becomes just one of many oppressions, with no particular centrality to social analysis (Wright: 1986, 117).

The approach in this paper takes as its point of departure the plurality of oppression *and* the relative primacy of class location in articulating through wider institutions, and thus selectively reinforcing, the other modes of oppression. The premise is that capitalist exploitation harnesses the existing sexual, caste or ethnic oppressions, that the hierarchic relations within the family and those within and between caste or ethnic communities mutate and adapt, instead of withering away, with technological modernization, in agriculture and industry.

The search for cheaper labour tends to maintain a labour force stratified by sex and by caste or ethnicity—with differential wages, conditions, and access to the means for better employment (these means include education, organization and networks of information and access, not just money and production assets). In the history of western industrialization, the use of unfree labour (indentured, loan-bonded or openly enslaved) was massive, as Charles and Louise Tilly (1981) note, in certain sectors of capitalist growth (in plantations, mines, railroad construction). In India, as also in other developing countries in recent times, including the East Asian success cases, economic growth has taken place within labour markets quite strongly segmented along the pre-existing social hierarchies of sex and caste or ethnicity, which have posed little hindrance to the exploitation of labour—defying elements of both neo-classical and Marxist positions.

This paper views class location, patriarchy, and the caste/ethnic hierarchy as interactive elements defining and differentiating the condition of women, channelling them into widely differentiated sets of work options and constraints. Sexual and other social hierarchies have remained aligned with inequalities of material assets, of education, and of organizational assets—inequalities that structure most resource allocation, including that by the state.

In this analysis, the interactions between class exploitation, social hierarchy and androcentric institutions (within family and outside) are seen as having the effects of (a) intensifying the oppression of women in the *lowest* socio-economic stratum; (b) undermining the process of class formation, to the extent women are in the labouring classes but without a commensurate share of organized union voice; and (c) diffusing the emergence in the public arena of female-focused goals and lobbies cutting across all the strata.

THE STRUCTURAL UNITY OF THE DIFFERENCES

In a country where tens of millions of women work as wage-labourers, millions as salaried regular employees, public and private, and where for hundreds of millions the daily workload differs inordinately, depending on private assets and access to basic public facilities, women's experience of oppression is not confined to the domestic arena of intra-family allocations or exchanges, but spans social relations and labour transactions as well. These three arenas are so closely linked that it will be incorrect to pick one — the androcentric institutions within family—and regard it as an undifferentiated, common denominator of women's oppression throughout the society.

The hold of patriarchal norms varies significantly across the class-caste strata, across regions and communities within India. Landless labour families (communities), often with neolocal or uxorilocal marriage customs, do not generally follow the highly sex-segregated work and living norms that the propertied, patrilineal, patrilocal families do. This difference is correlated with the former belonging mostly to tribal or outcaste groups and the latter mostly to upper caste or privileged ethnic groups. Persistent asset poverty goes hand in hand with bottom location in the social hierarchy.

Together, they continue to produce two things: (a) greater exploitation of women of the asset-poor labouring classes; and

(b) less drudge work, more time, and better opportunities of education and employment for women of affluent families than what would have been the case had social hierarchy and asset inequality not combined so mutually reinforcively. This does not imply absence of economic mobility, up or down. The upwardly mobile, with increased access to means and options, promptly adopt the norms observed by the social higher-ups, and educate girls for upwardly mobile marriage and bring educated brides, both having potential access to white-collar jobs if needed or desired. That, in fact, has started generating the supply of middle-educated young women increasingly entering the lower echelons of government jobs and also the temporary jobs in the recently expanding non-traditional industries intensive in semi-skilled female labour. However, their proportion is as yet too small to change much the strong positive correlation between asset poverty, bottom location in social hierarchy, and concentration of female labour in the lowest-wage, harshest-condition, and most denigrated sectors of casual wage-labour.

In this convergence of class differentiation and caste/ethnic hierarchy, the affluent social higher-ups traditionally have a two-sided reaction. The relatively unrestricted outside movement of labourer women of the tribal and outcaste communities, in large part imposed by economic hardship, is invariably viewed by those with somewhat higher social rank and economic level as something (a) to be conspicuously avoided by their own women; and (b) exploitable, sexually *and* economically, by men whose own wives and sisters are safely secluded or located in privileged job sectors. Toleration of sexual exploitation of labouring-class women in India is an integral part of the hierarchy of domination being maintained. Traditionally, men of subordinate class/caste/ethnicity have been, and to some extent still are, demeaned and broken by conspicuously denying them exclusive access to and control over their women's sexuality as well as labour power. As this control is perceived and practised as the privilege of the wealthy upper caste/ethnic strata, it is oppressively maintained in those strata *and* oppressively denied to those below. This is the connecting link in sexual oppression between the social strata, despite differences in forms and agents. There are also significant parallels and linkages between the oppression of women and that of outcastes and tribals in India. (Parallels have also been noted in colonial and racial subjugation.) It thus does not seem very logical to regard the oppression of women as unique, unparalleled, undifferentiated.

Capitalist profitmaking, it is increasingly evident, thrives on, at least is unhampered by, a segmented labour market hooked up with its cultural receptors in the existing hierarchies of sexual-cum caste/ethnic domination. The latter helps surplus extraction directly by funnelling a greater proportion of female workforce into low-wage casual labour in the informal substrata of capitalist enterprises, national or multinational. It also helps indirectly by having the female and juvenile poor shoulder more of the work of subsistence farming (and other non-capitalist sectors like domestic service), thus subsidizing the labour supply of male peasant migrants (and of lower middle-class women with some education to the capitalist sectors and their informal substrata). The social construction of the differences in women's work, and in related conditions including gender norms and relations, is far from a purely cultural matter, but is functional to, and indeed an integral part of, exploitation in basically capitalistic production relations.

The inescapable fact is that women in India, as also in most of South Asia, do not constitute a class in terms of either exploitation or shared interests. The nature of extra-economic oppression, too, differs by class location and is exercised by men as well as women. In the middle and upper socio-economic strata, women's oppression is largely generated within the family (especially the patrilocal extended family in village exogamy, isolating young wives from natal community). But in the lower strata, women's oppression comes mainly from the exercise of the socio-economic power hierarchy. Women from the upper and the lower strata in South Asia are not very likely ever to be united on the kind of 'speak bitterness' platform against husbands and mothers-in-law that occurred in the relatively undifferentiated peasantry of China in the course of its revolutionary movement. Nor are they likely to be united across the economic gulfs to demand priority for socialized childcare the way it occurred in socialist countries without cheap female labour for domestic service. Even with a common form of oppression, viz. male ingestion of alcohol, producing beating of women and drain of income, it is generally concealed less and intervened in more in the case of the labouring classes.[1] Women in general may be suffering more oppression than men, but it differs in nature and extent by their class location, and it is exercised by men as well as by women within and above their class. Not surprisingly, the staunchest female opposition to any solidarity movement of educated middle-class young women with labouring-class women comes not only from orthodox older women but also from the upwardly mobile among the lower social strata.

Paradoxical though it may seem, while women in India are divided by class and caste, their experiential differences continue to be interdependent, bound together in the hierarchic norms defining and maintaining their differentiated conditions. The ideology underlying the structure of experiential differences of Indian women divided deeply by class and caste/ethnicity/religion is now challenged in two ways. Women in the labouring classes are increasingly, through self-organization, confronting and resisting their exploitation. And by directly involving middle-class women (as activists, advocates, and lobbyists), they are also pulling feminist ideas and goals away from their earlier middle-class perspectives.

Recent experience shows that the issue of oppression of poor women by the socially powerful can be turned into a rallying point for uniting poor and middle-class women, increasingly aware of the structural link between the modes of their own oppression and the modes of oppression of women of the lower class–caste strata. In a society with persistent wealth inequality and ingrained hierarchy, but with universal suffrage delivered by national independence, the greatest potential for a solidarist feminist movement seems to lie in the issues most relevant for the bottom stratum, and, in areas affected by civil war, on the issues of women as victims, grievously affected directly and indirectly by police and military repression.[2]

I will now move on to some empirical details of the structure of differences in women's experience in India. I will specifically take up three things: (a) household asset level (land and non-land assets or means of production) as a determinant of the quality and quantity of women's work, the aggegation of which shows up as the class-gender interaction effect observable at the macro level; (b) the inter-relationship between class location, education and work pattern; and (c) the inter-relationship between class location, work quantity and quality, and fertility behaviour.

HOUSEHOLD ASSET OWNERSHIP AND FEMALE EMPLOYMENT QUALITY

In rural India, the percentage of households owning no land ($\leq$ .005 acre) decreased over the sixties (from 11.68 in 1961-62 to 9.64 in 1971–72) and increased over the seventies (to 11.33 in 1982).[3] Over these two decades, although the landless *proportion* has remained steady, their absolute *number* has increased by at least 2.5 million, i.e. by nearly a third. The near-landless (owning .005-.5 acre) proportion of all rural households increased from 32.5 in 1961–62 to 37 per cent in 1982; their absolute number increased by about 11 million, i.e. by nearly one half between 1961-62 and 1982—partly from inheritance subdivision and partly from regional success of land reform in creating micro owners. The majority of these two groups of rural households depend primarily on wage-labour for subsistence.[4] Some of those owning more than 0.5 acres, especially those located in arid areas of poor yield, and those who lost the use of own land mortgaged for loan, also depend mostly on wage labour. Altogether, four out of every ten rural households in India in 1983 were for the most part labour households, comprising over a third of rural population, four-fifths of them working mainly in agriculture.[5] India's rural proletariat, sub-proletariat rather, is far larger in size than the urban industrial proletariat. And this is even more so in the case of the female workforce.

Over a quarter of *all* rural households in 1982 had agricultural labour as principal livelihood.[6] Of the households owning zero to less than one acre (nearly half of all rural households), two-fifths were mostly agricultural labourers.[7] Leased area is quite small relative to owned area; and the vast numbers of the land-poor no longer account for much of the leased-in land.[8] The traditional tenancy option whereby the semi-landless worked in the past as rent-paying peasants has withered with the rise of agrarian capitalism.

A striking feature of the land poverty in rural India is its concentration in the disadvantaged (scheduled) caste and tribal communities. In 1982 these communities accounted for 14 per cent of all rural households owning no land (≤ .0005 acre), 46 per cent of households in the marginal owner (< 1 acre) category, and 17 per cent in the small owner category (1–2.49 acres). Other rural households constituted 9 per cent of the landless, 30 of the marginal owners, and 18 of the small owners.[9] Altogether, the scheduled castes and tribes accounted for about 30 per cent of all rural households in 1982, but only 18 per cent of all household-owned land.

Consistent with this, rural labour households are proportionally more from outcaste and tribal communities. In general, female wage-labourers in rural India are proportionally even more from these groups than male wage-labourers. The higher incidence of casual wage-labour in the case of land-poor tribal and outcaste women indicates the compounded effects of the three factors (viz. asset poverty, low social status, and sex) which structure the access to employment options and accentuate their exploitability in both high-growth and low-growth environments, though in different ways and to different extents.

The distribution of land (ownership and use) broadly indicates the class map of rural households in terms of the main productive asset. It shows the increase in recent decades in the ranks of landless and semi-landless labour households, labourers hired on casual, irregular basis at extremely low wages, mostly in agriculture, generally non-unionized, unlike the employees in organized private and public sectors, including the large tea and coffee plantations. The land-ownership pattern also shows the proportionally greater land-poverty of the scheduled castes and tribes, and the close alignment of social and economic inequality.

However, the asset location of *households* does not symmetrically show up in the work quantity and quality of male and female members. That depends also on the customs ruling intra-household division of labour and on two aspects of the household's economic environment—the development of local infrastructure and the environment of the common property resources that the land-poor depend on for a great many essential needs (or public-facility substitutes meeting those needs). Most (cross-sectional) data on the relation of individual employment (quantity and quality) to asset distribution (land, non-land) and to the regional development level broadly support this. The lower the household asset level, and the poorer the economic environment in which the asset-poor household is located, the heavier the extent of female workload and the worse the quality of the totality of their livelihood work (in terms of wages and conditions of the paid work, productivity in the case of the self-employed, and return to the energy spent in subsistence processing and gathering work). This holds far more strongly in the female than in the male case. Correspondingly, affluence shows up in sharper female withdrawal from wage-labour and hard manual work in field and home, for which the poorer women are often hired or equipment used.

It is well-recognized that even when there is a spurt in rural opportunities for regular salaried employment, for increasing productivity in self-employment, and for acquiring marketable skills, each of these is much less accessible to households in the lower socio-economic strata. The sex gap in the access to better employment, productive inputs, and education is greater and more persistent in the lower socio-economic strata, and the reason for this is not worse sexism in the labouring classes, a notion sometimes asserted, not established. The reason, rather, is the interlinked working of the institutions of social patriarchy and labour-market disadvantage. Another reason is the excessive total workload of the female poor, which is connected with their poverty of options and marginalization from productive resources, aside from the generational persistence of their educational gap.

Household asset poverty[10] and depressed local economic environment both increase the workload and decrease the quality of the work of the female poor. Sex-disparity in employment quantity and quality is evident from several indicators at various levels of aggregation. I will briefly note the class-and-gender compounded effect on the quantity and the quality of work.

First, in both rural and urban India, regular salaried employees constitute a much smaller proportion of the female than the male workforce. Casual labourers at the lowest wages with little job security or regularity constitute a much larger proportion of the female workforce.[11]

Second, across the states (Table A.1), greater incidence of rural landlessness (LL) tends to be accompanied by a significantly greater proportion of the rural female working population engaged in casual wage-labour (CL), as in Tamil Nadu, Kerala, Karnataka, and Maharashtra. The main exceptions are West Bengal, with high LL but low incidence of rural female CL relative to both the national average and the state-level rural male CL, and Andhra Pradesh, with about an average degree of landlessness but a comparatively high female incidence of CL. The main explanation is that while the landless in rural Andhra Pradesh are mostly from low castes and tribes,[12] extremely poor and concentrated in seasonal wage-labour in the state's intensive rice cultivation, the landless in rural West Bengal include many from higher-status communities having better access to education and a great deal of pressure to save face among social peers by secluding their women from field labour, which traditionally carried great social stigma in eastern India.

Third, the widest sex gap is in the proportion working in regular wage or salary earning jobs (Table A.1). Two things are involved here. First, although the sex gap in education is narrower in urban India, and decreasing quite rapidly in recent years, many educated women in the lower-middle classes continue to be housewives, despite the economic pressure to earn, in order to deal with the task of not just child rearing, harder in their mostly nuclear households despite fewer children, but the rather crucial upward-mobility strategy of coaching the children to excel at school and compete for the limited supply of elite jobs. In rural areas, prosperous farmers' daughters are increasingly schooled, but the chief motivation in their case, as against their sons', is upwardly mobile marriage to educated salary earners rather than to become salary earners themselves, although that too is occurring in some regions where suitable job sectors are coming up.

The proportion of women among regular employees has increased in recent decades, due to the expansion both of the pool of educated women and of the public sectors of white-collar and professional jobs. However, a far greater proportion of female workers remains in casual or irregular low-wage jobs: three out of every ten female workers in urban India in 1983, as against one out of every sixteen male workers; one out of every three female workers in rural India, as against one out of every four male workers.

On the one hand, the greatly increased entry of women in higher education and in the skilled workforce in recent decades can be seen as an emerging force of social change in sex roles. On the other hand, the enormous increases in the poverty-driven supply of non-literate female and juvenile labour in the proliferating sectors of low-wage casual jobs can be seen as a major source of irreversible change in traditional patterns and norms of the sexual divisions of labour that have prevailed with the household enterprise as the main mode of employment.

The significance of the effects of these two ongoing changes—the modernizing pulls that have brought middle-class women into salaried and professional jobs, political and administrative offices, and the distress that has pushed the female poor, adult and juvenile from subsistence production work (even if that was under patriarchal control) in the ranks of the proto-proletariat in the informal substrata and periphery of the formal capitalist sectors—may not yet be fully reflected in the prevalent social construction of gender. But one can reasonably regard this as a matter of time lag rather than disconnection of social ideology from the material conditions that differentiate people by work and define their relations in production.

The ideological control of gender roles and women's status is thus slowly, but surely, being eroded by the increasing presence of women both in the elite job sectors and in the harshly exploited informal labour sectors, neither of which is directly under the control of family patriarchy. In both cases control continues to be exercised indirectly through ideology, as indicated by the still almost universal acceptance of the traditional marriage norm (a woman has to be married once, and not below her natal family's social rank) both among the educated affluent and the uneducated poor. However, this disjuncture between the locus of female labour use and the norms of appropriate female role and behaviour may not justifiably be projected well into the future. Lagged transformation of gender roles occurred historically in other cases. Also, the case studies of women in either category in India, while showing the persistence of male-dominant ideology, also show increasing expression of the women's resentment of and impatience with that ideology (Gulati, 1982; Kapur, 1970; Mies, 1980).

Apart from the changing pattern and location of work, another set of factors that is bound sooner or later to undermine or at least modify the ideology of patriarchal domination is demographic. The population explosion of the sixties and the seventies has now produced a greatly enlarged ratio of younger to older persons, and this ratio will continue to enlarge for at least a couple of decades more. Because older women have crucial socializing and monitoring roles in the patriarchal control systems in South Asia, this unprecedented shift in the age distribution of population does constitute, together with the rising incidence of family nucleation both in the newly proletarianized sectors and in the educated middle classes, serious constraints for implementing the traditional norms. The population explosion has inevitably started weakening the age-and-sex hierarchy of patriarchal domination within the extended family and close community.

Corresponding to the disproportionately high female concentration in the ultra-exploited casual-wage sectors, and low female share of the regular employees, an important aspect of the class-gender interaction effect is the extent and content of women's work in household enterprises,[13] agricultural and non-agricultural. The proportion of female (family) self-employment (SE) is particularly large in three of India's states relative to male and relative to the national female average (Table A.1). These three states (Himachal Pradesh, Madhya Pradesh and Rajasthan) are relatively low both in agricultural growth and in the incidence of landlessness and semi-landlessness (LL + SLL), which have meant a smaller supply of wage-labour and greater scope for self-employment on land. These three states traditionally also have more small-scale animal husbandry, the work of which is largely done by women. In the southern and the eastern states (first three and last three rows of Table A.1), in contrast, the high incidence of rural landlessness has both lessened the scope for self-employment and pushed up the supply in the rural labour market, which affluent farmers have used to substitute the female family labour in farming and other manual work, unless the tasks are mechanized. The slightly above average level of household employment (SE), for both female and male, in the prosperous state of Punjab, in spite of the large farmers' affluence-prompted withdrawal of family women from farm work, is probably explained by the state's rapid growth of rural non-agricultural enterprise linked with its agricultural growth. In urban areas, across the states, the extent of women's self-employment (SE), relative to men's and relative to the national average, is high in south India where the female incidence of casual labour (CL) is *also* high, and in others (Rajasthan and Himachal Pradesh) where female CL is among the lowest. The extent of inverse correlation between female SE and CL observed at the rural state level, connected with interstate variation in the incidence of land-poverty, does not appear in the urban case. One reason for this is that part, an increasing part, of women's work in urban areas looks like self-employment but is really piece-wage home-based labour, at the bottom layer of subcontracts vertically integrated into big business. Far greater numbers of female wage-labourers are outside the trade unions, as a result partly of management strategy (of national companies as well as subsidiaries of the multinationals) and partly of the counter-productivity of the regulations supposed to benefit factory women.

Let me now move on to another aspect of the work regimen of the female poor. It concerns their average daily total workload in hours and physical energy spent; and it varies inversely with the household income level. The conflict between outside employment and the work of running home and rearing children is handled in the middle classes either by staying off the labour market or by hiring domestic servants and mobilizing help from the extended family. In the case of female labourers, including all those servant maids, desperate poverty resolves the conflict at the expense of their own need of family care, and at the expense of adult and juvenile females' barest minimum needs for rest, health, and education in the latter case. The endless workload of subsistence tasks and essential housework is particularly onerous for the female poor in rural areas, because of their poorer access to the basic utilities and equipment that could raise the productivity per unit of time and energy spent in the kinds of work that they must do every single day. The point, aptly summarized by Dixon-Mueller, is: 'The unequal distribution of resources *across* households, compounded by the sexual division of labour *within* households, produces a situation in which women (especially in asset-poor households) are relegated to activities with the lowest productivity and returns' (1985: 36).

The excessive loads of low-return work of rural females are aggravated also by depressed local economic environment, in which their greater seasonal unemployment imposes harder resort to the subsistence-gathering activities; secondly, by deforestation and depletion of the commons; and thirdly, by poor access to potable water, fodder, and cooking fuel, which have to be laboriously collected mostly by women and children.

The excessiveness of their daily workload, due to these factors, works to the detriment of (a) adult and juvenile female health, by producing caloric deficit and rest deprivation, and (b) girls' school attendance and completion, by leaving them too little time for it. The 'endless days' of physically exhausting labour in direct economic activities, in gathering and subsistence-processing tasks, and in housework (which takes enormous time and energy in the absence of easily available water-source, soap, and cooking fuel) produces not only the crushing quality of life of the female working poor, but also the generational reproduction of the education deprivation that perpetuates their worse poverty of work options. To emphasize this is not to deny that the uneducated poor are often unmotivated towards girls' education, but to stress that here is a concrete factor in the life of the working poor that public policy can remedy with certain targeted policy measures.

CLASS LOCATION, EDUCATION AND WOMEN'S WORK

In rural landless labour families in particular, the pressure of low-wage labour and the laborious gathering and scrounging for subsistence obstruct the schooling of girls, who take over or share the labourer mother's housework, childcare and gathering chores, as these families are often nuclear and lack the mutual help resource of the extended families of landholding peasants. Because the rural female wage-labourers are disproportionately from outcaste and tribal communities, the connection between women's load of low-wage work away from home and girls' loss of schooling is especially severe in their case, and reinforce their other problems related to their employment regimen.[14]

Taking education or marketable skills as an economic asset, a means to better work options, the appalling female poverty in education and its wider sex gap at the lower socio-economic levels is explained not only by the well-known factors of social discrimination and illiterate parents' indifference, but also by the heavier female loads of low-yield work. In order to enable the daughters of the rural poor to attend and complete at least primary, preferably secondary, schooling, the necessary set of steps in public policy is to improve their access to potable water and cooking fuel, to target part of social forestry on meeting the cooking fuel needs of the land-poor, and to generate through regionally suitable targeted programmes better slack-season employment options for women of rural land-poor families. Seasonal migration of adults, despite its contribution to survival and labour-market adjustment, absorbs large costs (private and social) in the loss of family-care, schooling, and important aspects of the quality of life.

In the state of Kerala, though poverty has not quite been eliminated, a high rate of relatively sex-balanced school enrolment has been achieved with better coverage of basic needs of the rural poor, enrolment drives, school lunch programme, free textbooks, and, not the least important, the massive politicization of the labouring classes. A similar process has started more recently in West Bengal, another state high in rural landless poverty.

In 1983, after nearly four decades of planning, economic and educational expansion, 80 per cent of rural India's female population aged 15 and above were illiterate, 14 per cent had some primary education, 6 per cent had middle to secondary education and 0.3 per cent were college graduates. Of the rural males 15 and over, 50 per cent were illiterate, 28 per cent had some primary level schooling, 20 per cent had middle-secondary schooling, and 1.4 per cent had college degrees (Table A.2). Of the urban females in 1983, 45 per cent were illiterate, 24 per cent had some primary education, 26 per cent middle-secondary schooling, and 4 per cent college education. Of the urban males, 20 per cent were illiterate, 28 per cent had some primary schooling, 42 per cent middle-secondary, and 9 per cent college education.

Clearly, the incidence of each level of education is much higher in the urban population and the sex gap much narrower. In both urban and rural areas, the sex gap is widest at the level of bare literacy, but decreases at successive levels of education. Table A.2 confirms what is well-known to casual observers, namely that the incidence of literacy and education level strongly correlates with the household income (expenditure) level. It shows two other things: at below the official poverty line, less than 1 out of every 10 rural females aged 15 or above and 2 out of every 10 urban females have primary education.[15] Even in the per capita expenditure group of Rs 300 and above (which at 1983 prices would roughly comprise the middle classes), the female incidence of illiteracy in 1983 is twice the male incidence and runs as high as 6 out of every 10 rural females aged 15 and over; and (d) that although the sex gap in the incidence of primary schooling is much lower above the Rs 200 expenditure level, the gap in primary-level completion remains much greater than in enrolment and continues to be very high in the middle-to-secondary levels in the rural population that is well above absolute poverty (Rs 100–300 per capita monthly expenditure).

A final point from Table A.2 is that the sex gap is the smallest among the middle-to-secondary educated and college graduates from urban middle-class households (with at least Rs 300 monthly per capita household expenditure). The pool of middle-educated women (nearly one out of every two in urban and one out of every five in rural India in the Rs 300 and above level) is large enough in absolute terms to hold no surprise about the recent growth of female entry into semi-skilled jobs in the white-collar sectors (social services and government offices) and in the expanding non-traditional private industries (pharmaceuticals, electronics assembly, software, export garments and so on). This growth, undoubtedly, will continue: both in the pools of middle-educated young women, and in those enterprises that can get in a position to tap those pools. The process has been promoted by macro policies through the eighties, giving liberal incentives to a range of modern small industries and export production sectors, many of which also happen to be intensive users of middle-educated, largely female labour employed partly directly but largely at arm's length on piece-wage homework contracts.

This new division of labour has come under attack from feminists, both Western and Indian, rightly concerned with the practices of temporary hiring, the active discouragement of trade unions, with the typical family strategies of investing the young females' temporary earnings in male-focused family goals, and with the apparent lack as yet of any progressive impact of this kind of employment on women's position in family and society and in the labour market. One must, however, point out certain things which these arguments do not take into account.

First, overstaffed as the government offices and services already are, the incidence of unemployment among the secondary and higher educated would have been much greater had the restrictive industrial policies of the sixties continued. Second, while the goal of universal primary education can, with proper social policies, be achieved irrespective of family income level and labour-market prospects, the stepping up of secondary school completion may be harder to bring about and sustain without economic incentive, i.e. without substantial growth in the semi-skilled job sectors. Third, even when a few years of a young woman's earning before or after her marriage is neither controlled by her nor a step toward her longer-term participation in productive labour, the experience of working and moving in the traditionally male-concentrated public domains predisposes the young towards modification of the repressive gender norms, even if it does not quite produce gender equality in work and in norms or ideology. Fourth, even a few years of earning before and/or after marriage may help slow fertility by raising the average female age at marriage and first birth and by spacing births. To the extent the incentive of a few years of such earning pushes up secondary school completion among rural poor teenage girls, it similarly may help the pace of fertility decline with a few years' postponement of marriage and childbearing.

CLASS LOCATION, FERTILITY AND WOMEN'S WORK

The association of lower fertility with higher incidence of women working outside the home is observed in the case of both educated middle-class families and agricultural labour families with neither land nor education, for very different reasons relating to their respective class location and to their differently caused high incidence of family nucleation. Educated, middle-class women tend to have higher marriage age and fewer pregnancies, and a substantial proportion of them now have access to well-paid jobs with leave and other benefits. Landless labourer women, with very little of either food intake or rest, have to work at extremely low-paid and strenuous jobs that offer neither leave as a matter of right nor regularity of earning. Their low fertility rate, compared to landholding farmers, is at least partly due to poor nutrition and health. The number of live children per agricultural labour household is even smaller because of the higher infant mortality in their case, due to extreme poverty, maternal overwork and the disintegration of the extended family network under pressures of landlessness, wage-labour, and migration (Krishnaji: 1989).

The main cause of higher infant mortality uncompensated by higher fertility in the landless rural families is the harshly exploitative work conditions of women as wage-labourers. Contrary to the notion that fewer children make for a better level of living and make women available for employment, fewer children in this case result from abysmal poverty and under-nutrition of the female working poor, from the health-ruining load of women's excessive labour, day in and day out. In this and other sub-proletarian labouring classes, child quality as well as quantity suffer, along with young women's health and quality of life, because of their poverty of work options and lack of access to marketable skills and means for self-employment.

The crushing combination of low-wage daily labour in harsh and relentlessly exacting conditions, and reproductive labour, with the lack of minimal social services and basic-need facilities, is reflected in the age-distribution of agricultural labourers. For instance, in 1977-78, of all female agricultural labourers aged 10 or more and belonging to rural labour households, 25 per cent were in the age-group 15-24 years and 56 per cent in the age-group 25-49 years, figures that are at least as high as the corresponding figures for males in the occupational category,[16] in sharp contrast with the usually low female rate of work participation in India in the prime reproductive age range.

This gender equality of the proto-proletariat—in relentless exploitation and acute deprivation—is as shocking as the gender equality of the educated urban affluent in salaried and professional jobs is impressive. Add to this the half-empty stomachs, the enormity of labour involved in getting water and firewood alone, the absence of paid leave or any surplus to tide over a day off work away from home, the hours of work in rain and sun, and the miles of walk to and from work. Then perhaps we begin to comprehend how ruinous the female workloads are in the vast ranks of the asset-poor in India.

As for the relationship between class location, the mode of female employment, and fertility behaviour, it is known that peasant families and others similarly deriving their main livelihood from household enterprises show a fertility pattern distinct from the asset-poor labour households on one side and the educated middle classes on the other. Peasant households, with female members working mostly in or near home, tend to have larger numbers of births and surviving children than landless labour families, because they usually have some food through the year, are less exposed to seasonal starvation, and because the home-based work of women and the availability of childcare resources within the extended families typical of the peasantry are more beneficial to infant and maternal health. All this is also more conducive to a consistent pursuit of their fertility goal, which is usually to have more sons as future workers on the farm and in other occupations, and as sources of old-age security, and as part of the patriarchal norms. That this is not quite the case with the low-end casual labourers—the construction and harvest trampers, and the hard-pressed migrants splintered off their kinship support—must concern public policy in order to ameliorate the harsh impact of women's work regime on, in particular, maternal and child health.

SOME CONCLUDING REMARKS

The main empirical points of this paper concern the wide difference between the social classes in women's education level, work quality and quantity, and work-related life conditions (health, fertility, total availability of family-care from household, community, and public sources). The differentiators in our reference to social classes are not just material wealth (land, money), but also skill assets (education level, which in large measure is a function of parental income and mothers' work conditions at low income levels) and organizational assets (e.g. access to union or guild-type organization, political voice in demanding and getting the needed basic services, access to mutual-help resources ranging from networks of contact and information for profitable migration strategy, to the care-giving resources generated within extended families and kin groups).

The main conceptual point of the paper concerns the ideological structure of interdependence that binds the differences (in oppressive conditions) between the classes into a rather resilient equilibrium. The three main sources of departure away from that equilibrium can be seen in the grassroots organization of women emerging in the lower strata, in the way this has involved sections of educated middle-class women as activists, and in the recent increases in a variety of government and NGO projects which have been attempting to combine elements of economic and social change in poorer women's conditions.

Economic exploitation of the asset-poor and their social oppression (by sex and by caste or ethnicity) interlink, overlap, and reinforce each other quite substantially. Women's access to economic opportunities is stratified by both family assets and personal assets (especially education and marketable skills, which are less randomly distributed — i.e. more systematically related to family income, parental aspiration, and social policy—than assets like prized beauty and useful cunning). It is also stratified by social hierarchy as the cumulative source of privilege and deprivation. Patriarchal family hierarchy is not uniformly oppressive across the socio-economic strata; women in the bottom strata generally experience oppression and exploitation from the upper strata more than from within their own family or community.

The social construction of the differences in women's experience, and in behavioural norms between social classes, is integral to the maintenance of the social relations of hierarchy and the class relations of exploitation, which are connected and interactive in both precapitalist and capitalist production relations. Women in India, and in much of South Asia, are divided by class (asset ownership), and by caste or ethnicity, and yet their different experiences are held together by the hierarchic norms defining and regulating their respective positions.

It would thus be incorrect to think of women as a class in terms of exploitation or as a community of shared interests and problems. The nature and the extent of oppression suffered by women differ according to their class locations, and in varying forms are imposed by men and women of the upper strata as well as by men within their own social class. For solidaristic feminism to reach across deep-rooted dividing lines, the essential question is about what kind of efforts are being made to ameliorate the condition of women located in the bottom strata, and to forge alliances for a broad-based social movement that takes explicit account of the class and caste/ethnic dimensions of most gender issues. For women labourers, the source of patriarchal oppression is more in the societal attitudes and institutional processes than in their own families. The zealous circumscribing of female sexuality in the upper social classes is linked with, indeed it is the flip side of, the sexual exploitation of lower-class women by upper-class men and its passive acceptance by many upper-class women.

Recognizing that for poorer women oppression comes mainly from the upper social classes and from the cumulation of socially hierarchic deprivation is crucial for the feminist movement to be able to cross these great divides. Recognizing that the differences in gender oppression in the different classes are really interdependent is central to conceptualizing the diversity of women's experience, and forging unity across the diversity and inequality. Of the two most promising developments in India since the mid-seventies, one is that more feminist-unionists, recognizing how collective self-organizations of women in low-income occupations can resist their usual demoralization and abjectness in the face of denigration, are involved in promoting and supporting their organization into guilds and unions. The other remarkable development is that more of the female working poor are becoming increasingly willing to work towards their own grassroots organizations to gain the socio-political voice that is indispensable for securing access to economic inputs and social services, and what is at least as important, securing protection from the harassment and extortions they constantly have to take from the higher-ups.

Stressing the differences in women's social experience has been viewed negatively by some feminists, but others have viewed it positively.[17] The perspective of this paper shares the second viewpoint. Recognizing the social and economic inequalities that exist among women is not divisive, but absolutely necessary for modifying those inequalities, at least for pressing towards informed policies to ameliorate the hardships generated by the inequalities for those at the low end of the socio-economic scale.

At the core of the hierarchy of social classes in the conservative social construction of female diversity, the hierarchic values and behaviour norms for women belonging to the different socio-economic strata. The forces of economic and technological change, opportunities for education and new employment, and the growing politicization and self-organization of the oppressed groups exert some modifying effects. However, the existing power hierarchy constantly applies sanctions and rewards to conserve the ideology of differential female conditions. The challenging task in this context, it seems, is a progressive social construction of the differences in source of oppression for women in different social classes, and the idea that is central to this is that the different forms of subordination are indeed interdependent. Feminism can work as a major counterforce by questioning the differential norms at the core of the socio-economic hierarchy.

NOTES

1. In Manipur and Tripura poorer women even banded together to attack liquor dealers operating in their community and to get the latter to chastise the offending husbands.
2. To mention just two examples, the organizations of mothers of killed and disappeared youths in Sri Lanka, and the sustained involvement of educated women and students in helping the mostly female and juvenile Sikh refugees of the Delhi killings following Indira Gandhi's assassination.
3. From 17th, 26th, and 37th rounds of the National Sample Survey. These and the other figures in this paragraph are from *Sarvekshana*, 11(2): 7–9, October 1987.
4. I say the majority, not all or most, because, some of the rural households owning zero to less than 0.5 acre derive the major part

of annual household income either from salaried non-manual jobs or from self-employment in owned plus leased-in land or in non-farm enterprise.
5. These figures are from the 38th round of NSS. *Sarvekshana*, 9(4): 11 April 1988.
6. About one-half were self-employed in agriculture; and about a quarter had non-agricultural livelihoods (as self-employed, as employees, or as employers or rent and interest receivers).
7. From Tables 2 (S–32) and 3 (S–46) from *Sarvekshana*, 9(2), October 1987.
   The wage-dependent percentage drops sharply with a little greater landholding, because they can lease in land more easily than those poorer in landownership. Of those owning 1–2.49 acres (nearly a fifth of all rural households in 1982) only 14 per cent lived mainly from agricultural labour, 70 per cent on farm self-employment, and 16 per cent on other income sources.
8. All leased-in land in 1982 averaged only 7.5 per cent of all owned land. The virtually landless (owning under .05 acre) accounted for only a fifth of total leased-in land. The small-to-medium owners (2.5–9.99 acres) accounted for one-third, and the larger owners (10 acres and above) for one-fifth. The lessees in rural India are no longer typically petty owners. Medium to large owners lease parcels to and from each other for convenience in machinery use, from small owners unable to obtain irrigation and other inputs, and by using the mortgaged land of near-landless debtors. These figures are from Tables 1 (S–18) and 6 (S–102) of 37th round NSS (source: *Sarvekshana*, 11(2), October 1987).
9. From Tables 1 (S18), 8 (S130), NSS, 37th round, *Sarvekshana*, 11(2), October 1987).
10. The individual, in alienable asset level of education and marketable skills, depends, especially for the female, upon the household asset level.
11. According to the 1983 National Sample Survey, the distribution of principal-status workers (age 5 and above) by the mode of employment differs by sex as shown below, by per cent.

	self-employed	regular employees	casual employees
Rural			
Female	54.1	3.7	42.2
Male	59.5	10.6	29.9
Urban			
Female	37.3	31.8	31.0
Male	40.2	44.5	15.3

SOURCE: *Sarvekshana*, 11(4): 39, April 1988.

12. 32 per cent of rural households in Andhra Pradesh and 37 per cent in West Bengal were from the scheduled castes and tribes in 1982. But, while in Andhra Pradesh they accounted for only 17 per cent of household ownership of land, in West Bengal they accounted for 25 per cent.
13. Whether as self-employed or as unremunerated helpers in family enterprise, comprising both agricultural and non-agricultural activities.
14. In labour households from scheduled castes/tribes in rural West Bengal, the percentage of girls 8–14 years old attending school in 1972–3 was less than one-fifth of that in upper-caste households farming more than 2.5 acres, while the percentage doing mostly domestic chores was twice as large in the former (K. Bardhan, 1984, 139–40). In between these two socio-economic groups, the percentage of girls attending school varied positively with farm size and caste status, and the percentage occupied in domestic work and gathering tasks varied negatively.
15. Urban location makes some difference even for the poor, but not that much really. Local school availability, cultural environment, and juvenile load of chores are collateral determinants.
16. 32nd round NSS data. *Sarvekshana*, April 1987, S–5. Of the male agricultural labourers belonging to rural labour households in 1977–78, 24 per cent were aged 15–24 and 55 per cent aged 25–49.
17. In the context of America's multi-ethnic, multi-racial as well as multi-class society, for instance, the need to address and incorporate rather than underplay the implications of these major social differences and corresponding differences in their primary concerns has been stressed.

TABLE A.1
RURAL LANDLESSNESS AND DISTRIBUTION OF FEMALES AND MALES (5+ YRS)*
BY EMPLOYMENT CATEGORY, 1983

	Landless and Semi-landless			% Distribution of Rural Population (5+)									% Distribution of Urban Population (5+)							
				Female				Males				Female				Males			
	LL	S-LL	Σ	CL	SE	RE	Σ	CL	SE	RE	Σ	CL	SE	RE	Σ	CL	SL	RE	Σ
Tamil Nadu	19.2	47.9	67.1	25	23	2	51	28	30	9	67	8	11	5	24	14	21	26	60
Kerala	12.7	63.5	76.2	12	19	4	35	23	23	9	55	5	14	6	25	19	19	17	55
Andhra Pradesh	11.9	37.4	49.3	27	25	2	54	25	37	8	71	7	11	4	22	9	24	26	59
Karnataka	13.7	26.0	39.7	22	23	1	46	25	39	4	68	8	7	5	21	12	20	26	58
Gujarat	16.8	23.9	40.7	17	26	1	44	20	36	5	61	4	7	4	15	9	21	28	58
Maharashtra	21.2	21.5	42.7	25	26	1	52	22	34	9	64	6	6	5	17	7	19	32	58
Madhya Pradesh	14.4	20.2	34.6	16	34	2	52	16	44	6	67	6	7	4	17	9	23	25	57
Rajasthan	8.1	14.1	22.2	6	48	0.5	54	8	50	5	63	2	17	2	21	6	29	19	53
Himachal Pradesh	7.7	18.8	26.5	0.5	51	1	52.5	7	50	6	63	1.5	11	7	20	5	21	35	60

	Landless and Semi-landless			% Distribution of Rural Population (5+)									% Distribution of Urban Population (5+)							
				Female				Males				Female				Males			
	LL	S-LL	Σ	CL	SE	RE	Σ	CL	SE	RE	Σ	CL	SE	RE	Σ	CL	SL	RE	Σ
Punjab	6.4	51.1	57.5	3	27	1	31	14	44	8	66	2	9	7	18	6	28	26	60
Haryana	6.1	39.6	45.7	4	25	1.5	31	9	39	9	57	0.5	8	2.5	11	7	30	24	61
Uttar Pradesh	4.9	38.7	43.6	5	25	0.3	30	11	47	4	62	2	6	4	12	6	31	20	57
Bihar	4.1	50.8	54.9	12	15	0.4	27	21	33	4.5	59	4	7	3	14	9	27	21	56
Orissa	7.7	32.4	40.1	14	20	1	35	22	36	7	65	5	3	3	12	9	19	30	58
West Bengal	17.2	47.0	64.2	7	14	1	22	23	30	8	61	3	6	6	18	7	22	30	59
All India	11.3	36.9	48.2	13	24	1	39	18	38	6	63	5	8	5	17	9	23	26	58

NOTES:
* Classification of workers in the table is for usual-status workers, both main and marginal.
LL: Landless households owning no land or less than 0.005 acre (as homesite, for example).
S-LL: Semi-landless households owning 0.005–1.0 acre.
CL: Casual-wage labourers in agricultural and non-agricultural work.
SE: Self-employed plus unpaid helpers in household enterprise, agricultural and non-agricultural.
RE: Regular employees on wages or salaries.
Σ : CL + SE + RE. This is the worker-poulation ratio in population aged 5+ years, by sex and location.

SOURCE: From NSS 38th round data. *Sarvekshana*, 9(4): S–125 to S–169, April 1986.

TABLE A.2
DISTRIBUTION OF PERSONS (15 AND ABOVE) BY EDUCATION AND SEX: HOUSEHOLD EXPENDITURE GROUPS

Rural India: 1983

Per capita Household expenditure (Rs/month)	Females (15 years and above)				Males (15 years and above)			
	Illiterate	Upto Primary	Middle–Secondary	Graduate and above	Illiterate	Upto Primary	Middle–Secondary	Graduate and above
30–40	93	6.6	0.9	–	75	19	5	0.1
40–50	93	5.7	1.0	–	72	21	7	0.3
50–60	92	6.3	1.4	0.02	70	22	8	0.2
60–70	90	8.1	2.1	0.01	63	25	12	0.3
70–80	86	10.7	2.4	0.05	56	29	14	0.5
90–100	83	13.0	4.4	0.10	51	30	18	0.7
100–125	78	15.4	6.8	0.18	45	32	22	1.3
125–150	72	18.2	8.9	0.38	40	31	27	1.8
150–200	68	19.7	11.5	0.67	37	31	29	2.4
200–250	61	22.5	15.3	0.75	33	29	34	4.1
250–300	60	21.5	16.9	1.59	30	28	38	4.7
300 and above	58	21.4	18.7	1.89	30	27	37	6.3
All groups	80	13.6	6.3	0.29	50	28	20	1.4

Urban India: 1983

Per capita Household expenditure (Rs/month)	Females (15 years and above)				Males (15 years and above)			
	Illiterate	Upto Primary	Middle–Secondary	Graduate and above	Illiterate	Upto Primary	Middle–Secondary	Graduate and above
30–40	76	15	9	—	48	33	18	1.6
40–50	75	14	11	0.2	48	30	21	1.7
50–60	74	18	8	0.2	48	30	21	0.8
60–70	70	20	9	0.3	41	32	26	1.2
70–80	65	22	13	0.4	36	36	26	1.9
90–100	58	25	15	0.8	28	36	33	2.6
100–125	52	26	21	1.7	23	33	40	4.5
125–150	44	28	26	2.2	18	32	43	5.9
150–200	38	26	32	4.2	14	27	49	10.0
200–250	29	25	39	6.8	12	22	52	13.8
250–300	24	23	44	9.4	9	20	53	17.5
300 and above	19	18	45	18.0	9	15	51	25.7
All groups	45	24	26	4.4	20	28	42	9.4

SOURCE: NSS 38th round; *Sarvekshana*, 11(4): S–22 to S–24, April 1988.

NOTE: Per capita monthly expenditure group Rs 0–30, a destitution level, is dropped here.

BIBLIOGRAPHY

Kalpana Bardhan, 'Women's Work, Status and Welfare: Forces of Tradition and Change in India', *South Asia Bulletin*, Spring, 6(1), 1986.

——, 'Work Pattern and Social Differentiation among Rural Women in West Bengal', in *Contractual Arrangements, Employment, and Wages in Rural Labour Markets in Asia*, edited by M. Rosenzweig and H. Binswanger, New Haven: Yale University Press, 1984.

Zarina Bhatty, 'Muslim Women in Uttar Pradesh: Social Mobility and Directions of Change', in A. de Souza, ed., *Women in Contemporary India*, Delhi: Manohar Publications, 1975.

Patricia Caplan, *Class and Gender in India: Women and their Organizations in a South Indian city*, London: Tavistock Publications, 1985.

John Caldwell, 'Mass Education as a Determinant of the Timing of Fertility Decline', *Population and Development Review*, 6(2), 1980.

John Caldwell, P. Reddy and Pat Caldwell, *The Causes of Demographic Change: Experimental Research in South India*, Madison: University of Wisconsin Press, 1988.

Bonnie Thorton Dill, 'Race, Class and Gender: Prospects for an All-Inclusive Sisterhood', *Feminist Studies*, Spring, 1983.

Ruth Dixon-Mueller, *Women's Work in Third World Agriculture: Concepts and Indicators*, Geneva: International Labour Office, 1985.

Leela Gulati, *Profiles in Female Poverty*, London: Pergamon Press, 1982.

ILO, *Yearbook of Labour Statistics*, Geneva, 1977–1985.

Pramila Kapur, *Marriage and Working Women in India*, New Delhi: Vikas Publications, 1982.

N. Krishnaji, 'The Size and Structure of Agricultural Labour Households in India', in Gerry Rodgers, ed., *Population Growth and Poverty in Rural South Asia*, New Delhi: Sage Publications, 1989.

Joanna Liddle & Rama Joshi, *Daughters of Independence—Gender, Caste and Class in India*, New Delhi: Kali for Women, 1986.

Maria Mies, *Indian Women and Patriarchy: Conflicts and Dilemmas of Students and Working Women*, New Delhi: Concept Publishing Co., 1980.

Joan Mencher, 'Peasants and Agricultural Labourers: An Analytical Assessment of the Issues Involved in their Organizing', in

P. Bardhan and T.N. Srinivasan, ed., *Rural Poverty in South Asia*, New York: Columbia University Press, 1988.

Asok Mitra, *India's Population: Aspects of Quality and Control*, Vol. I, New Delhi: Abhinav Publications, 1978.

Sarvekshana, Journal of the National Sample Survey Organization, Dept of Statistics, Ministry of Planning, Government of India, 8–11, 1984–1988.

Ursula Sharma, *Women, Work and Property in North-West India*, London: Tavistock, 1980.

Andrea Menefee Singh and Anita Kelles-Viitanen, *Invisible Hands: Women in Home Based Production*, New Delhi: Sage Publications, 1987.

Charles and Louise Tilly, *Class, Conflict and Collective Action*, Beverly Hills: Sage Publications, 1981.

Erik Olin Wright, 'What is Middle about the Middle Class?' in John Roemer, ed., *Analytical Marxism: studies in Marxism and Social Theory*, Cambridge: Cambridge University Press, 1986, pp. 114–140.

———, *Classes*, London: Verso (New Left Books), 1985.

6

Patriarchy and the Process of Agricultural Intensification in South India

PRITI RAMAMURTHY

INTRODUCTION

This paper focuses on how the process of agricultural intensification has affected women's work and lives in a semi-arid region of South India.[1] It explores the hypothesis that agricultural intensification, in this case stimulated by the introduction of a canal system and Green Revolution technologies, has led to more work for women both outside and within the home. It argues further that the returns to women from this increased burden of work, and the mechanisms by which output increases are distributed, are determined by a system of capitalist patriarchy; both gender and class differences are therefore significant.

In recent years several important theoretical and empirical studies on the gender and class implications of agricultural modernization in India have become available (for example, Agarwal, 1988; Sen, 1982). Most of these studies have drawn on data from the core Green Revolution belts—the mono-crop rice and wheat growing areas in the north and south—where the effects of mechanization on women's work have been particularly deleterious. This paper, by contrast, provides empirical information from a region which adopts a diversified, labour-intensive cropping pattern, with important implications for the sexual division of labour and the capitalist relations of production. Second, the existing data base on women's work lacks detailed studies such as Mencher and Saradamoni's (1982) on 'wet' rice cultivation for the hitherto 'dry' areas of South India. These are necessary because as Attwood (1988) and Ludden (1988), among others (Stein, 1980; Beals, 1974; Beteille, 1974) have emphasized, contrasting changes are occurring in the historically 'wet' and 'dry' zones. But none of these authors have focused on the gender implications of such changes. This paper, based on fieldwork in villages at the tail end of a canal system in upland Andhra Pradesh, attempts that focus.

GENDER AND AGRICULTURAL INTENSIFICATION

According to modernization theory, intensification through technological change in agriculture (for example, the introduction of irrigation and the Green Revolution package of seeds, fertilizer, etc.) is supposed to lead to the following general benefits:
(1) an increase in agricultural output through (a) an increase in yields of existing crops; (b) an increase in the gross cropped area or the bringing of land previously not cultivated under cultivation; (c) an increase in the cropping intensity or the number of crops planted on a given plot of land; and (d) a shift to a higher yielding strain of crop.
(2) an increase in the demand for labour time through (a) an increase in gross or net cultivated area; (b) an increase in yields and therefore harvesting operations; (c) an increase in the overall care and supervision necessitated by the new varieties; and (d) an increase in the number and kind of agricultural operations as a result of a change in the cropping pattern (Mellor, 1976; Dhavan, 1988).
As a consequence of the increase in output, the income of

cultivating families is expected to rise and as a result of the increased demand for labour time and increasing real wages, the income of agricultural labour families is supposed to rise (Agarwal, 1981).

However, as Ester Boserup (1970) first emphasized, gender is a basic factor in the division of labour and there are constraints on *women* realizing these benefits. Coming from within the modernization perspective, Boserup links the marginalization of women workers in the process of development to the fact that the economic gains that accrue to them are less than those that accrue to men. But as Beneria and Sen (1981) point out in their critique of Boserup, this ignores the historical-colonial and current processes of capital accumulation, and the effects of these processes on technological change and women's work, and on women as members of different classes. The general argument of feminist scholars, such as Beneria and Sen (1981), is that capital accumulation separates direct producers from the means of production and thereby makes their survival more insecure and contingent. This tendency manifests itself in new forms of class stratification and can have a variety of effects on women's work, depending on the specific form of capital accumulation in a given region.

Following from a critique of modernization theory, the theoretical premise of this paper (broadly based on Beneria and Sen, 1981; Sen, 1982; and Mies, 1986) is that gender differences in the impact of agricultural intensification are linked to two aspects of women's subordination. First, women belong to households that differ in access to land, other means of production, and wage incomes. As a consequence the conditions of work for women—in cultivating households on family farms or in the home, and as agricultural labourers—depend on the survival strategies of households in specific relation to the means of production (land, cattle, etc.) and to rural power relations.

Second, women are members of rural households, and their relationships with other members of these hierarchical social units embody relations of domination or subordination based on gender and age. The specific form that family relations take may vary according to class and caste.

While both rural power relations and family power structure provide the basis for different forms of women's subordination they are reproduced by an overall ideology that plays a critical role in the social construction of gender and in the process of women's subordination (Agarwal, 1988; Mies, 1986). Thus, women are defined primarily as dependent beings and are denied control over property, income, and active participation in public, 'political' decision-making. Patriarchal dominance within the home is reinforced through the sex-based division of labour and male control of women's reproductive capacity.

AGRICULTURAL INTENSIFICATION IN THE DRY-ZONES

Ethno-historians such as Ludden (1988) and Attwood (1988) have emphasized the social, economic and historical differences between two agro-technological regimes in India—the 'wet' and the 'dry'. Wet zones are areas with assured rainfall and drainage irrigation, where cultivation has centred on rice for many centuries. Rice cultivation has provided a secure subsistence base and required heavy inputs of labour. Wet zones are thus characterized by high population densities with rigid, polarized stratification systems: small numbers of high-caste non-cultivating landlords have traditionally dominated large populations of low-caste landless labourers. In contrast, cultivation in the dry zones (technically this includes semi-arid zones), with sparse and uncertain rainfall concentrated during a few months of the year, has centred on millet and sorghum. These crops have low labour demands and provide an insecure subsistence base. Dry zones are thus characterized by lower population densities with more flexible stratification systems dominated by 'yeoman peasants' from the middle and upper-middle levels of the caste hierarchy, smaller populations of low-caste landless peasants and fewer non-cultivating landlords.

Obviously, this classification into wet and dry zones is not meant to be applied rigidly or to exclude transitional regions of the subcontinent. Its importance lies in the fact that it allows us to start grouping fairly distinctive systemic characteristics so as to analyse the interaction of these with the process of agricultural intensification. According to Attwood (1988: 15), '(e)cologically and economically, the most favoured agricultural areas today are those portions of the dry regions which have received large-scale irrigation systems during the last century or so'. It is in these areas that irrigation reduces the risk of subsistence crises and significantly multiplies the return to cultivators from their land. The newly-irrigated land can support a denser population and small holdings become more profitable. Since these areas start with lower population densities, the pressure to expand the production of traditional subsistence crops is less; the switch to high-value cash crops is therefore possible. Through this process both small holders and landless labourers are expected to gain more in economic opportunities than they lose. In addition, since prevailing stratification systems are relatively more flexible, agricultural entrepreneurial castes are able to gain access to land, labour and capital (Attwood, 1988).

But, as mentioned in the introduction, the authors of such historical ecological analysis have so far been silent about the relationship of wet and dry systems to the sexual division of labour. Others, such as Mencher and Saradamoni (1982) in their comprehensive study of rice cultivation, have documented that in the highly stratified wet areas it is the women, especially tribals and those of the lowest castes, who have been involved in all aspects of rice cultivation (except ploughing). Although in their study of rice cultivation in Kerala, Tamil Nadu and West Bengal they found differences in the specific tasks women perform and 'suspect that both caste and ecological factors play some role in the development of diverse patterns', in general they observe that in the past the introduction of more intensive agricultural technologies such as animal-powered ploughs did not reduce the disproportionate amount of work done by women (Mencher and Saradamoni, 1982: A, 163).

Other researchers, such as Bardhan (1981) and Miller (1981) have suggested that the lower rates of female labour participation in the wheat-growing areas of the north—or the dry zones—can be linked to a greater degree of discrimination against women. Based on secondary sources of information on Haryana and Punjab (also in the dry zone), Sen (1981) suggests that the proportion of women in landless labour households is lower in those subregions which have adopted the new technologies most heavily. But the growth in the number of small farms and the 'finance intensity' of the new technology has led to greater participation of women from small landowning households in the labour force.

Unlike in the north, female labour participation rates have always been high in the south, even in the dry zones. Agarwal (1988) relates regional differences in the sexual division of labour to cultural norms relating to female seclusion and control over female sexuality. Women have played and continue to play a significant role in rainfed cultivation. It is the purpose of this paper to illustrate that the recent process of intensification in the hitherto dry zones in the south has put a disproportionate burden on the women. While the contribution of women is apparent and 'visible', it is not necessarily socially recognized as valuable (Agarwal, 1988).

Based on the theoretical considerations outlined in this and the previous section, my main task in the rest of this paper will be to demonstrate the effect agricultural intensification and the process of capital accumulation have had on the sexual division of labour and on women's workloads in agricultural labour, marginal, middle and rich landowning households. I will also examine how women's control over the labour process and its product has changed, and explore how the process of agricultural intensification has affected pre-existing patriarchal controls within the household. My concern throughout will be to show how the system of capitalist patriarchy uses existing gender hierarchies to keep women in subordinate positions at each level of interplay of gender and class in a hitherto dry zone in south India.

THE RESEARCH LOCALE

The south-eastern region of Andhra Pradesh, or the uplands (in contrast to the fertile, long irrigated deltas), are characterized by low (29" annually) and uncertain rainfall, 77 per cent of which falls during the five months of the south-west monsoon. Consequently, the area is dry and drought-prone. The one large-scale canal system built in the region by the British proved to be so expensive and such a poor yielder of revenue that it was not until the post-Independence period that irrigation works were embarked on to protect the area from repeated droughts.

The research locale examined here is fed by a canal that flows off the Tungabhadra river in south India. The scheme was built in 1956 as a famine relief system, with the short-term purpose of providing employment through its construction and the long-term goals of providing drinking water and enough irrigation to meet subsistence food requirements in as many villages as possible. Although at its inception cultivators were reluctant to use canal water except in years of drought because of the higher risks and costs entailed, by the early 1970s the availability of canal water made it possible to adopt the Green Revolution package of HYV seeds, fertilizer, etc. Since then, water is stolen by head-enders to intensify agricultural production, depriving aggrieved tail-enders of their authorized share.[2]

Water is provided to the cultivators in the 80 villages fed by the canal, which is 89 miles long, in either the first (June-December) or the second (December-April) season. In the first season of 1984-5 there were 15,000 acres of paddy, 8000 acres of cotton, and 1000 acres of vegetable crops. In the second season, other than the cotton (which is a 7-9 month crop), there were 21,000 acres of groundnut, and 2500 acres of paddy. In addition, in the first season, about 40,000 acres of *jowar* (millet) and 30,000 acres of tobacco, cotton, groundnut, etc. were grown with monsoon rainfall.

In 1984, the total population of the *taluk* (administrative subregion) through which the canal flows was 152,000. Of the 25,000 cultivators, 6000 were women. The number of agricultural labourers was 40,000, of whom 22,000 were women (Census of India, 1984). The sex ratio (or the number of females per 1000 males) was 975 for Andhra Pradesh, fourth highest in the country. When compared to sex ratios of 879 in Punjab and 870 in Haryana, this bears out the expectation that high female labour participation rates are associated with low gender disparities in child survival (Agarwal, 1988).

The distribution of land in each village community is highly skewed: 5 per cent of all landowners are magnates or rich cultivators holding 20–200 acres of land each; they own 25 per cent of the land; 25 per cent of landowners are middle-income cultivators holding 5–20 acres each; they own 40 per cent of the land; 70 per cent of all owners are small holders or marginal cultivators holding less than 5 acres each; they own 35 per cent of the land; 54 per cent of the population owns no land.

The caste classification of the population broadly parallels class or landholding classification. The magnates are of the 'higher' or upper-middle Reddy or Kamma castes. The Reddys have been the dominant caste in the region for many generations. They are the old elite: numerically small but owners of a disproportionate amount of land, they wield enormous economic, political and social power. The Kammas migrated into the region from the delta area of Andhra Pradesh in the mid-1960s. They number only in the thousands but are the new entrepreneurs: they bought irrigated land cheap from uninitiated native cultivators soon after the introduction of the canal, and cultivate high-value crops for profit.[3] Middle and marginal cultivators are of the 'middle' and service castes (Kurruva, Boya, Golla, Muslim, etc.) and agricultural labour are from the 'lowest' and untouchable castes (Mala, Madiga, Teliga, etc.). Brahmins or the 'highest' caste are numerically and politically insignificant in the region. Culturally, the Reddys or 'warrior-peasants' and Kammas or 'commercial peasants' are closely involved in the day-to-day operation of their holdings, unlike their counterparts in the wet areas 'whose efforts to increase profits ... tend to focus on problems of social control: how to extract more, where possible, from labour' (Ludden, 1988: 111).

Because the canal was designed as a 'protective' rather than a 'productive' system, but is currently used for the express purpose of agricultural intensification, canal water is a scarce commodity. Villages at the tail-end of the system are at a disadvantage because water must first flow through head-end communities. Tail-end villages must organize collectively to mobilize water or they receive none. This is of relevance because only women who live in tail-end villages, which do have autonomous water associations, are directly involved in the process of agricultural intensification.

The research focused on three villages at the tail-end of the canal system, all of which had water associations, and one village at the head-end. In each village, magnates, middle and marginal cultivators and landless labourers were interviewed. Within each household, both women and men at various stages of the life cycle were questioned.

The empirical findings which follow concentrate on the work of women as workers outside of the home and link this to their subordination within the home and the ideology of gender. While I recognize the importance of detailed information on and analysis of female reproduction, unfortunately this was beyond the scope of my initial study.

TRANSFORMATIONS IN THE SEXUAL DIVISION OF LABOUR AND WOMEN'S WORKLOADS

A. *The Technical Organization of Labour*

'All our lives are a summer time.' This statement by Maryamma, a woman and an agricultural labourer, sums up the close dependency of finding employment on the availability of water; during the dry summer months there is very little work available. In the days before canal irrigation the only source of moisture was rainfall, and this is still the case for the majority of plots in the area. A comparison of the technical organization of labour under rainfed and irrigated conditions demonstrates the process of transformation.

The rainfed cycle of agricultural production usually begins in June-July and ends in December-January, except for the minor work of tobacco curing which goes on till March. Preparation of the land for sowing can begin only after the rains (early June), as the black cotton soils in the area cannot be worked unless first moistened. As is the practice in all other parts of the country, only male labour is used for ploughing, though women may assist in clearing the fields and manuring. Working the bullock team and seed drill to sow groundnut and jowar is also an exclusively male job; one woman may follow the team dispensing seed. Rainfed cotton, however, is sown by women as each seed is individually implanted. Women prepare tobacco nurseries and transplant the seedlings: these are nearly exclusively women's jobs, as is the process of de-budding to allow for greater leaf growth. De-budding tobacco is a particularly offensive task which leaves a blackish poison on the hands; women usually miss their mid-day meal when doing this work so as not to ingest the poison. Harvesting the leaves and stringing tobacco garlands are also considered female jobs. Both men and women work on curing the leaves for the next three to four months.

After planting there is little work for women on cotton, groundnut and jowar until the harvest, as it is the men who are involved in interculture—hoeing and fertilizing. If the rains are good and weed growth excessive, women may be employed once for weeding. Cotton is picked using women's labour, with the number of individuals hired varying with the quantity of cotton to be picked. *Gumpu*s or groups of women and a few men usually take responsibility for harvesting groundnut and jowar. For groundnut, harvesting entails uprooting the plant, plucking the pods off stems, and binding the stems. Women are again employed to clean the nuts and bag them. For jowar, harvesting entails reaping, cutting the cobs off the stems, and binding the sheaves. Both men and women are employed to thresh and winnow the grain.

The irrigated agricultural production cycle starts with the release of water into the canal by the authorities of the Irrigation Department.[4] The main irrigated crop in the first season is transplanted, HYV paddy. Ploughing of the paddy fields, building and repairing bunds are all male tasks; levelling work is done by women. Nurseries are tended for twenty-five days, after which women's labour is used exclusively for transplanting. Men procure and apply water, broadcast fertilizer, and spray pesticide. Weeding, which is done at least twice, is women's work. Harvesting operations—reaping and binding, followed by threshing and winnowing—are carried out by groups of women and men.

The next irrigated crop to be sown is hybrid cotton of both the long- and short-duration varieties. In addition to the agricultural operations necessary to grow rainfed cotton, women work on hand-weeding (after the men hoe), and basal fertilizer application three or four times during the crop cycle. Harvesting employs approximately three times the number of woman-days required to pick rainfed cotton. The most labour-intensive crop to be grown is hybrid cotton for seed. Though it is grown on less than 5 per cent of the irrigated land, it employs mainly children almost continuously for 4–5 months to cross-fertilize each flower of one variety with the pollen of another.

The most significant irrigated vegetable crop is onion. Again, men plough the fields, build the furrows, and irrigate the plots. Women transplant the seedlings in the furrows. Weeding is done two or three times during the crop cycle by the women and involves a lot of work, as no hoeing is possible. Harvesting—

uprooting the bulbs, trimming the stems, and bagging—is done in groups of both men and women.

Irrigated groundnut is the major crop in the second season. In addition to the agricultural operations necessary to grow the rainfed crop, women shell twice the quantity of seeds, and weed once or twice following hoeing by the men. Harvesting by groups employs at least twice the number of women-days required by rainfed groundnut.

B. Women's Workloads

Table 1 summarizes the labour demanded to grow the different crops. The transformation in women's work as a result of irrigation is dramatic. In contrast to the 25 woman-days for jowar, paddy requires 53 woman-days per acre. Compared to 44 days for rainfed cotton, irrigated cotton requires 112 woman-days per acre, and compared to 23 days for rainfed groundnut, irrigated groundnut requires 45 woman-days per acre.

More generally, changes in the technical organization of labour as a result of agricultural intensification mean that even marginal cultivators must hire labour for transplanting and harvesting. Moreover, the specific cropping pattern adopted in the area, historically and currently, has not lent itself to heavy mechanization,[5] with subsequent unemployment for women. Instead, there has been an increase in the overall demand for labour and a greater increase in the availability of work for women because the sexual division of labour is such that the more arduous, labour-intensive jobs have been and continue to be done by women.

A greater increase in women's jobs compared to men's was also found in the west zones by Harriss (1982) for the North Arcot region of Tamil Nadu, and by Mencher and Saradamoni (1982) for Chingleput, Tamilnadu and Kuttanad, Kerala. In contrast, in the dry zones of Punjab and Haryana, Sen (1982) argues that the 'mechanization' effect has caused a narrowing of the tasks women labourers do. In the rice-growing Thanjavur district of Tamil

Nadu, Gough (1977 and 1982) found that the growth of a relatively large surplus of agricultural labourers led to men being hired for work traditionally done by women. *There have been no comparable changes in definitions of the sexual division of labour yet in the region studied.*

Unlike mono-crop rice or wheat areas, there is a diversified cropping pattern in the region of Andhra Pradesh studied. Hence, seasonal peaks and troughs in employment opportunities are not as marked. Women agricultural labourers have a higher chance of finding some employment over a longer period of time as the various crop cycles overlap and require a wide variety of agricultural operations. However, this means that women must be skilled at several operations for many crops.

WOMEN'S CONTROL OVER THE LABOUR PROCESS

Connections between agricultural intensification, capital accumulation and women's work become more apparent when analysing wage and labour relations. In the hierarchy of labourers—permanent, regular casual, and seasonal casual—women are at the lower end (Sen, 1982). Unlike men who may be employed as *ghasaghadlu* or attached labourers for a year, women can never be so certain of continuous employment.[6] It is inconceivable that they should be employed as attached labourers because that involves practically full-time work and would leave them with no time to meet the needs of their families. Thus, the ideology of gender, which assigns primary responsibility for the work within the household to women, has a direct effect on their participation in work outside the house, in the sphere of production.

For generations women have worked as agricultural labour or coolies for the families of rich and middle cultivators. The women refer to their work as *kashtam* or hardship. This form of coolie work and overt production relations continues to this day, but the essence of these production relations has undergone significant changes (Mies, 1988). For example, according to Ramlamma, an elderly woman whose husband has been a ghasaghadlu for more than thirty years, in the past 'practically the whole village' used to be employed by the biggest landowning family in one of the sample villages. All those employed had the right to do certain jobs for which they would receive remuneration, usually in kind. Today only about fifty women are employed by the family during the agricultural season as regular casual workers or those who are 'free to work for others if [their] permanent employer has no work' (Mencher and Saradamoni 1982). Of these fifty women, only about ten retain a right to doing *inti panni* (work around the house) for the family when there is no work in the fields (*polala panni*) during the summer or monsoon. The remaining forty women are in a position where neither traditional relations exist nor are they free wage labourers. The magnate can be fairly sure of his labour supply during the agricultural cycle and he need not be burdened by a wage bill during the slack months. But the labourers are in no position to bargain for adequate wages to compensate for those off seasons.

The chances of a woman being a regular casual labourer are higher if she is married or related to a male attached labourer, because recruitment of labour by large- and middle-sized cultivators is through their attached labour. Thus, within agricultural labourer households, there are those who are more disadvantaged than others; without contacts to a rich or middle landowner through male attached labour, the chances of women finding employment in slack periods are low. Sen (1982) makes a similar argument for Thanjavur where, with the decline of *pannaiyal* (bonded) relations, a woman is no longer able to ensure a steady subsistence for herself and her children through her connection to the male member of the household.

The most frequent form of employment is seasonal casual work for which individual women may receive a daily wage or a piece rate. For some operations, such as transplanting or harvesting, women may receive a share of the amount her *gumpu* or group has contracted. The daily wage rates or equivalents paid per person per day in 1984–5 are presented in Table 2.

From the Table it is apparent that the highest wages are received for the group activities of transplanting and harvesting, where in a narrow window of time a lot of work must be accomplished. Gumpus have varying numbers of people, depending on the size of the plot to be worked. Usually the group consists of related households and rarely cuts across caste lines. Both men and women form groups, and women are paid the same rates as men (reported also by Gough, 1977; Harriss, 1982; and Mencher and Saradamoni, 1982).

Women are paid a lower wage for all other tasks where the sexual division of labour is explicit. During the summer and other periods of slack employment, the daily wage rate for women drops to a minimum of Rs 3.00. At other times it fluctuates around Rs 5.00. This is in comparison to a male wage rate of Rs 4.00–5.00 off season and Rs 8.00–10.00 during the season. The justification men give for women being paid lower wages for exclusively women's jobs is that these jobs need less strength. The justification is based on the assumption that women are weaker than men ('can they lift boulders?') and that women cannot do as much as work as men, i.e. that their productivity is lower. But when I confronted the men with the fact that women worked as hard as they did and were paid as much during harvesting, for example, the men fell back on 'tradition' as a legitimate basis for women's lower wages ('from the time of our ancestors that is the way things have been').

Piece rates are paid for such jobs as plucking groundnuts, shelling groundnuts for seeds, stringing tobacco garlands, etc. A woman's 'productivity' is gauged by how much work she has been able to accomplish, regardless of the time it has taken. Such a system of measuring work means women spend long hours doing intensive and arduous tasks. Yet it is these very jobs that are defined as 'light' because they are done in the landowner's home or on the village threshing floor, not in the fields. The social stigma of doing 'light' work has been effective in keeping the number of men doing these jobs low; and this is in the interests of the landowners who would have to pay higher piece rates if men were employed.

Since the intensification of agriculture, all wages are paid only in cash except those for the harvesting of food grains—paddy and jowar. Women workers feel that wages paid in grain are more valuable than wages paid in cash because they can be stored and used to meet household consumption needs, whereas cash is eaten away by rising prices. Both Harriss (1982) and Frankel (1971) also found that the payment of wages in kind was a hedge against inflation. Frankel even argues that the discontinuance of wages in kind may have contributed to the development of agrarian unrest in Thanjavur.

When there is an increase in demand for labour during the rains or harvest season, the money wage-rate increases. However, this does not mean that wages are determined only by the market forces of demand and supply. Although wages are not decided by the village panchayats (councils), there seems to be tacit agreement among the *peddamanshulu*, or Big Men, as to how much should be paid. The average daily wage paid to men and women hovers around the minimum necessary for family subsistence. While seasonal casual labourers, especially those from other villages, may be tempted away from the magnates for a few days by a slightly higher wage, regular casual labourers are aware that only their *sahukar* or patron may provide employment during slack periods. Labourers and magnates explain this phenomenon in patronage terms, but the reality is that landowners require a stable workforce and a constant supply of labour. In a less guarded moment one of the largest landowners admitted, 'I employ these women not to do [them] service. Since I am in the village I am looking after my own development'. That labourers find security of livelihood in paternalistic relations with cultivators has been noted by Harriss (1982) and Epstein (1967) as well.

The demand for labour during peak periods is partially met by increasing the length of the working day. It is important to note that women with infants cannot join the groups because they cannot leave their children alone from 8 a.m. to 7 p.m.; for this reason they are also less likely to earn sufficient money from piece work, where the longer you work the more you can earn. Peak demands for labour are also met by using immigrant labourers from unirrigated regions; these labourers arrive in large groups and camp in the fields for the entire duration of the harvest season, thereby dampening potential increase in wages due to temporary scarcity.

Women agricultural labourers are employed for approximately 20 days a month from August to April, for a total of about 180 days per annum. This is less than the 310 days that Harriss (1982) reported for North Arcot, but considerably more than the 90 days estimated by Djurfeldt and Lindberg (1975) for Chingleput, Tamil Nadu, or the 80 days estimated by Epstein (1962) for Mandya, Karnataka. But, as Mencher and Saradamoni (1982) point out, the number of days women work varies even within the same village depending on such factors as whether or not they are regular casual workers, their age, marital status, husband's contribution, etc.

To recapitulate, with the introduction of irrigation and the intensification and diversification of agriculture, the demand for women's labour has increased. But the discrimination against women continues in the form of lower wages or exploitative piece rates for all exclusively female jobs. This is possible because it upholds and is derived from a patriarchal vision of the male as the primary breadwinner and the male wage as the main source of family income. The fact that women's wages are viewed as supplementary explains why they are willing to work for lower wages than men. Dependence on men is further reinforced through labour relations; women who are connected to male attached labourers have higher chances of securing employment. Paternalistic relationships with cultivators increase the security of workers' livelihoods and the chances of being able to borrow in contingencies. The system of patriarchy is thus independent but closely linked to the structure of class exploitation. It has a material base. It enables relationships of exploitation between the classes to be upheld by allowing the payment of inadequate wages to both men and women. As a result, while rich landowners are able to milk larger profits, agricultural labour families are left with nothing after everyday subsistence needs are met. This increases their dependence on the rich, to whom they must turn for loans.

TRANSFORMATIONS IN THE SEXUAL DIVISION OF LABOUR, WOMEN'S WORK, AND THE LABOUR PROCESS IN CULTIVATOR HOUSEHOLDS

Following from the theoretical premises outlined earlier, it could be hypothesized that with the intensification of agricultural production, the work of women of marginal cultivator households on their own farms increases and precludes their working on other farms; and that the work of women as supervisors in middle and large cultivator households increases. Or alternatively, that patriarchal proscriptions are re-shaped and women withdrawn from fieldwork as the new culture makes women's work invisible by restricting it to the home, while simultaneously endowing men, whose wives are now defined as dependent housewives, with greater prestige (Mies, 1986).

In fact, as we found in the previous section, with the introduction of capitalist agriculture the technical organization of labour has changed so that even the smallest cultivator needs to hire non-family labour at certain times in the crop cycle. To minimize the cost of hired labour—no longer paid in kind but in cash wages—family labour is used as much as possible. Moreover, the financial intensity of irrigated agriculture—the need to purchase seed, fertilizer, pesticides, etc.—forces women from marginal households to work on other farms as well, to generate cash incomes. While the men of these households busy themselves with land preparation, grain processing and selling, during the dry season the women migrate to irrigated villages to work as wage labourers. Thus again, because the ideology of gender views women's earnings as supplementary and defines certain jobs as 'light' and female work, it is the women who are pushed out to find work and are exploited by those who own substantially larger landholdings and who can pay them less than they would if they employed male labour.

There is a tendency for women of households with large landholdings to disengage from direct work but this seems to be less important for a man's status and prestige than Epstein (1962) found in Mandya and is more comparable to North Arcot (Harriss, 1982). Except for the Komati and Kamma castes (less than 5 per cent of all cultivators) there was little evidence here of men bragging that their womenfolk do not even know how to transplant or weed (Epstein, 1962).

The reason why patriarchal norms have not been redefined in upper-middle cultivator households can be related to the fact that a major source of power for the men in these households is derived from their participation in alliances with other men—in the village peddamanshulu (Big Men) council (which substitutes for the moribund statutory village panchayat), in government offices, and in the extended family.[7] The building and maintenance of these exclusively male alliances takes a great deal of time and resources, and therefore the women cannot be withdrawn from the field. In fact they play a critical role in the supervision of day-to-day agricultural operations. This entails managing the labour force, and seeing they are provided with supplies such as seeds for sowing or fertilizer. In addition, women cultivators often join in the field activity to set the pace for the labourers and substitute as hired hands. As they work in the field, alongside other women, these women correct their work as it goes along, and make on-the-spot decisions about what to do next. In contrast, when men go to the fields to supervise work, they do not actually soil their hands or do women's work, as that is culturally unacceptable. Only women in the very rich households are not involved in supervision these days, although they were just two generations ago; a male *ghumasta* or supervisor is hired, in addition to employing many attached labourers.

The case of women from the Kamma caste is noteworthy because although they form less than 5 per cent of the population they are aspiring for the position of a new rural elite. Patriarchal domination by the Kamma men limits women to work within the household—caring for the cattle, and feeding the labourers.[8] Kamma men are highly critical of the local men and criticize them for being 'lazy' and letting their womenfolk work in the fields; they question the very manhood of these men. Whether Kamma traditions and social norms will be imitated by the rural populations that are dependent on them remains to be seen. Currently the economic, political and social control of the old elites and the material necessity for their women to be out working in the fields allow their norms to dominate.

Women of marginal landowning households receive no compensation for working on their own lands. When working as casual daily wage labourers, their contribution ensures that minimum family livelihood needs are met. Women of larger landowning households are also not compensated for their work either as supervisors or as labourers. It was not possible to compute how much these women contribute to family income, or how the fields in which the women worked differed from those where women did not, but clearly without their work it would be impossible for their families to maintain their class position.

Women contribute to family subsistence in other ways as well; this work is often subsumed as 'supplementary' work and rendered of less importance than it is. For example, women of landless and marginal families grow nurseries for tobacco, onion, and other vegetable crops. This activity is carried out during the slack months of May-June. Nurseries are sown near the village on communal caste plots and watered from wells or streams. After the rains, seedlings are sold to cultivators at a modest profit. Often marginal and middle cultivator households plant highly labour intensive garden crops—tomatoes, chillies, beans, etc.—in a portion of their plots, and the women manage the entire picking operation with family labour. This is used for family consumption and sold on the local market. These crops yield quick and regular returns. The care of small animals—chickens, goats, and cattle—is also the women's responsibility.

To recapitulate, the transformation of agricultural production has led to changes in the technical organization of labour that necessitates the hiring of labour even by marginal households. The purchase of labour and HYV seeds, chemical fertilizers and pesticides instead of their organic substitutes entails a higher investment outlay. As these households do not have investible surpluses, they are forced to borrow. Often this is from the magnates at usurious rates of interest (24–60 per cent per annum). Even if the borrowings are from the nationalised banks, the financial intensity of the new mode of production forces the women of these households out into the wage labour market. For the same reasons—labour requirements and financial intensity—and the fact that their menfolk are busy with public, political roles, the women of middle households are not withdrawing from agricultural production. They not only supervise the labour force but join in with their own labour. Only the wives of the magnates have withdrawn from direct agricultural production. In a region where patriarchal domination did not include proscriptions on women's participation outside the home, the economic pressures of agricultural intensification have fallen heavily on women, and consequently there has been little reshaping of pre-existing patriarchal norms.

WOMEN'S WORK, CONTRIBUTION AND CONTROL WITHIN THE HOME AND IN THE PUBLIC ARENA

A woman's typical day begins at around 4.30 a.m. and ends at around 10.00 p.m. During this time she is responsible for looking after the children, cooking, cleaning, animal husbandry, fetching water and firewood, and working for a wage or on the family holding. When the male is an attached labourer, there is no chance that he will help with any of these tasks as he leaves the house early in the morning, returns to eat and leaves again to return late at night. When the male is also a casual wage labourer or a marginal cultivator, the women said that they were lucky if the man helped them fetch water and looked after the cattle. Not only does a woman do all this work, but her own needs are always attended to last and often insufficiently, reflecting her position of subordination to the men, and male children.

The roots of women's oppression can be traced to their domestic work and the interaction between production and reproduction. Although detailed information on reproduction is not presented, from field observations it is clear that women have primary responsibility for reproduction and maintenance of the future labour force. In addition to childbearing and child rearing, women bear the burden of all (or most) of the domestic work within the house. As Mies (1986) has pointed out, this work plays an important role in the economic system but it is unrecognized and unpaid. It provides commodity labour power and this, as already mentioned, enables both men and women to be paid less than necessary to finance household reproduction and maintenance. A corollary of this system is that single women and households headed by widows are the worst off.

In addition to the contribution of women in the sheer physical effort of maintaining a family, they are often forced out of the house to feed hungry children. Mencher and Saradamoni (1982: A165) estimate that for landless households 'the ratio of female to male contributions is at least 1:1 and in most cases it is higher', despite the lower wages that women get. Even in marginal households, they estimate that women contribute 'a little under half to well over half of household income'. These conclusions were borne out by the Andhra Pradesh case material as well.

Despite the significant contribution of women to household income, given the overall ideology of gender it is the men who control its allocation. In the family of Yellamma, a labourer of the

Madiga (Harijan) caste, her husband and his brother are attached labourers, and her parents-in-law and daughter are non-earning members of the family. Yellamma gives her entire wage to her husband and so does her brother-in-law. The husband gives his mother Rs 20 every day to meet daily expenses. The rice, jowar, dal, cooking oil, salt, and chillies that are consumed daily cost about Rs 22. Once or twice a week, 'as the menfolk's mood takes them', they may give an additional Rs 3 to 6 for vegetables or meat. Yellamma or her mother-in-law sometimes ask the men for this extra money. Expenses such as clothes, house repairs, cooking vessels, etc. are heavy, and if they cannot be made at the time the men receive their wages every four months, the men borrow money from their employers. They also borrow from their employers at the time of large expenses like weddings or illnesses.

Narsamma and her husband own 1¾ acres of dry land on which they grow jowar, and 1½ acres of irrigated land on which they grow groundnut. In 1984–5 the rains failed and the one quintal of jowar they have harvested was only sufficient to feed themselves, their four children and his mother for a few months. The groundnut brought in a profit of Rs 2600; after liquidating the crop loan of Rs 1500, there was only Rs 1100 left. As this was insufficient to meet family needs, Narsamma works as a agricultural labourer whenever she is not working on her own fields. Her husband works as a community water guard for four months a year. And their eldest daughter works for the village magnate cross-fertilizing cotton. Again it is the man who allocates the money for daily expenses and gives additional money for vegetables about once a week and for meat about once a month.

Both Yellamma and Narsamma do not question the right of the men to assert control over the cash income. According to Yellamma 'we cannot depend on our uncertain earnings to raise a family'. Narsamma concurs, adding: 'it is the man of the house who can get the loans'. Yet clearly without the women's income neither of these families would be able to subsist. Male control over incomes is even greater in the households of middle and large cultivators where the women do not earn a cash wage. Moreover, among all classes and castes, some part of the men's salary is spent on personal consumption: tea, bidis, toddy or other alcoholic beverages, movies, gambling, and womanizing. This spending is taken for granted as a normal male prerogative by most women, even though it means that household earnings are diminished (see also Mencher and Saradamoni, 1982). Further, this personal consumption by men has increased with greater cash incomes. According to one male informant, the quantity of alcohol consumed in his village has increased four-fold in the last fifteen years.

Although the women of middle cultivator households bear so much of the burden of agricultural activity and the men acknowledge that without them the productivity of the labour force would be very low, they are affronted with the idea that their wives and mothers take agricultural production decisions. 'What do they know? They are not educated. They do not move around in public or discuss the latest hybrid seed or pesticide with others. Besides the burden of making those kinds of decisions and the commensurate risk is all on our heads', said one male informant.

The devaluation of women's contribution and their powerlessness has been traced to the exclusive male control over land in a situation of extreme scarcity of land amongst the peasantry (Kishwar, 1984). As in other areas of the country, in this region too nearly all land titles are in the names of men. At the time of land ceiling legislation, land belonging to the richest landowners was sub-divided among family members and registered in the names of the women, but this is only a nominal transfer. In the few instances where land was actually given up, it was transferred to landless men, not women.

Despite their 'visibility' in the fields, women remain invisible in the public, 'political' domain. Women are conspicuous by their absence from all forms of collective activity, whether it is the sheep-penning auctions, the community court of village peddamanshulu or Big Men, the farmers' association, the water association or the temple board. Because women are prevented from forming such alliances at the village level, they are politically powerless (Kishwar, 1984). The wives or widows of magnates are the only ones that have ever run for political office and they have not been concerned about the condition of other women.

CONCLUSIONS

When asked whether agricultural intensification had benefited them, women agricultural labourers and marginal cultivators replied 'No'. Though they acknowledge that the demand for labour has increased, they said the wages they receive in cash get spent immediately on daily consumption, usually at the discretion of the men. In the days before canal irrigation, whatever both men and women earned as daily labourers or attached labourers was in kind; the grain could be stored for household consumption throughout the year. Women of marginal cultivator households complained that not only has their workload increased but so has the investment necessary to engage in agricultural production. As a result, they are forced to work as agricultural labourers on the lands of others, and the HYV rice they grow is sold while an inferior variety of rice is bought for household consumption. At the end of each season, loans are repaid with high rates of interest, leaving no investible surplus and the cycle of poverty continues.

In contrast, when men who are middle and large cultivators were asked the same question they said: 'Yes, not only have we benefited but agricultural labour has benefited too'. First, they pointed out that real wages of agricultural workers have increased. They calculate that before the canal the average male attached labourer got 5 quintals of jowar and Rs 500 in today's prices, or a total value of Rs 1800, whereas today the average attached labourer gets Rs 2800-4200. In fact, the issue of whether real wages have increased or decreased as a result of agricultural intensification is a highly debated one.[9] Second, the cultivators argued, agricultural labourers and marginal cultivators have shifted their consumption habits from 'inferior' grains like ragi or jowar to 'superior' grains like rice; and third, their disposable income has increased—tea and toddy, for example, were never consumed in such large quantities as nowadays. In fact, although rice is socially considered 'superior' to jowar, it is much less nutritious.[10] And, by their own admission, any increases in disposable income are spent by the men, usually on themselves, not on improving family livelihoods.

The diametrically opposite assessments of agricultural intensification are understandable given the fact that any increases in output and earnings are enjoyed disproportionately by landowners at the cost of the landless, by larger landowners at the cost of the marginal, and by men at the cost of women. This is because mechanisms of output distribution are upheld by both current processes of capital accumulation and the overall ideology of gender.

More specifically, I have tried to illustrate how in a hitherto dry region of Andhra Pradesh, intensification has not led to changes in the sexual division of labour. In the sphere of production, women are burdened with greater workloads; but the ideology of gender continues to define exclusively female work as light, allowing women to be paid lower cash wages than men, or exploitative piece rates. In marginal cultivator households, women are pushed out to earn wages as agricultural labourers. And in the households of middle cultivators, women are working as labourers and supervisors, forcing other women to toil harder. None of this work by landowning women is paid for or socially valued. Within the home, regardless of the actual contribution of women's earnings, it is the men who decide on its allocation. And although women are working more and harder outside the home, the burden of their work within the home has not changed. Not only do women have to perform the work of childbearing, but child rearing and maintaining the family too continue to be defined as women's work. Women are denied access to land and prevented from public, political participation; they are thus, in general, powerless.

In conclusion, favourable assessments of the impact of agricultural intensification in the hitherto dry regions of south India need to be informed by an analysis of gender-specific effects. I have tried to show here how processes of capital accumulation are significantly linked to the ideology of gender. Defining women as supplementary earners allows them, and men, to be paid lower wages as agricultural labourers than would be required for family survival. In addition, the system of large landowners employing male permanent labourers ensures the availability of a relatively captive and stable female labour force—the relations and kin of permanent labourers—even though these women are not employed during the slack periods. And the substitution of cash wages for foodgrains enables landowners to reap the profits from higher foodgrains prices, since wages have typically lagged behind prices (Agarwal, 1988). Thus the system of patriarchy upholds class exploitation and is linked to the development of agrarian capitalism in an essential way.

TABLE 1
WOMEN'S LABOUR TIME DEMANDED FOR MAJOR CROPS

Crop	Labour Demand (in woman-days per acre)
Rainfed jowar	25
Rainfed Tobacco	5
Rainfed Cotton	44
Irrigated Cotton	112
Irrigated Cotton Seed Production (mostly children)	1891
Rainfed Groundnut	23
Irrigated Groundnut	45
Irrigated Paddy	53
Irrigated Onion	125

TABLE 2
WAGE RATES OR EQUIVALENTS PER PERSON PER DAY FOR 1984–85

Month	Female/Male/Children/Group	Individual wage or equivalent in Rupees	Piece or contract rate in Rupees	Activity
May	Female	3.00	–	Cotton picking
	Female	4.00–5.00	1.25/dabba	Groundnut plucking[1]
	Male	4.50–5.00	–	Clearing fields
	Male group	4.00–5.00	1.00/yard	De-silting canals
June	Female	3.00	–	Levelling, manuring paddy fields
	Female	1.50–3.00	0.25/seer	Shelling groundnut for seeds[2]
	Male	5.00	–	Clearing fields
	Male	15.00	15.00/acre	Building canals & drains for paddy fields
July	Child	3.50–4.00	–	Cross-fertilizing cotton[3]
	Female	5.00	–	Transplanting onion, sowing groundnut & cotton
	Female group	10.00–14.00	100.00–140.00/acre	Transplanting paddy[4]
	Male	8.00–10.00	–	Ploughing paddy, sowing groundnut.
	Male group	10.00–20.00	10.00–15.00/acre	Irrigating cotton

Month	Female/Male/Children/Group	Individual wage or equivalent in Rupees	Piece or contract rate in Rupees	Activity
August	Female	4.00	—	Weeding paddy, onion, cotton, basal fertilizer for cotton
	Female	5.00	—	Tobacco transplanting
	Male	8.00–10.00	—	Fertilizer application, hoeing
September	Female	5.00	—	Weeding onion, tobacco transplanting, basal fertilizer for cotton, *jowar* sowing
	Male	8.00–10.00	—	*Jowar* sowing, hoeing, fertilizer application
October	Female	5.00	—	Cotton picking, weeding, fertilizing *jowar* sowing
	Male	8.00–10.00	—	*Jowar* sowing, hoeing, canal cleaning
November	Female	5.00	—	Cotton picking, tobacco de-budding
	Female	3.00–6.00	0.50/*seer*	Groundnut shelling
	Male & female group	18.00–22.00	150–180 *seers* paddy/acre	Harvesting paddy[5] (transplanted)
	Male & female group	9.60–12.00	48–60 *seers* + 30 *seers* paddy/acre	Harvesting, threshing (broadcast)

Month	Female/Male/Children/Group	Individual wage or equivalent in Rupees	Piece or contract rate in Rupees	Activity
December	Female	4.50–8.75	0.30–0.35/kg	Cotton picking[6]
	Female	6.00–12.00	0.30/dorram	Tobacco garlands[7]
	Male	8.00–10.00	–	Groundnut sowing, irrigating
	Male & female group	5.00–8.00	–	Onion harvesting
	Male & female group	6.00–7.00	60.00–70.00/acre	Groundnut harvesting
January	Female	5.00	–	Groundnut sowing, weeding, cotton picking
	Male	8.00	–	Groundnut sowing, irrigating
	Male & female group	6.75	54 seers jowar/acre	Harvesting *jowar*
February	Female	4.00–5.00	–	Groundnut weeding, cotton picking, tobacco curing
	Male group	10.00–30.00	10.00–14.00/acre/wetting	Irrigating
	Male & female group	6.00–12.00	9.00/tractor trip	De-silting canal

Month	Female/Male/Children/Group	Individual wage or equivalent in Rupees	Piece or contract rate in Rupees	Activity
March	Female	4.00–5.00	–	Tobacco curing, cotton picking
	Female	5.25–8.00	1.75–2.00/dabba	Groundnut plucking and cleaning
	Male & female group	7.00–9.00	140.00–180.00/acre	Groundnut uprooting acre and plucking
April	Female	5.00	–	Tobacco curing, cotton picking
	Female	6.00–11.00	2.00–2.75/dabba	Groundnut plucking cleaning
	Male	8.00–10.00	–	Transporting harvest
	Male & female group	10.00–13.00	200.00–260.00/acre	Groundnut uprooting and plucking

NOTES TO TABLE 2

1. A *dabba* is an empty oil tin used as a 'standard' measure. Average number of dabbas a woman can pluck a day is 3–4.
2. Average number of *seers* shelled per day is 6–12.
3. This activity employs children every day till January/March. However, less than 5 per cent of the gross cropped area was used for seed cotton production.

4. A group of 20 can transplant 2 acres per day.
5. A group of 20 can finish all harvesting operations except winnowing in a day. Conversion factors: After milling the rice yield from paddy is 66 per cent by weight. 1 *seer* = 0.93 kg. Open market price of rice in November 1984 was Rs 4.00/kg.
6. In most villages cotton picking is paid for as a daily wage. In one village a rate/kg was paid. On average a woman can pick 15–20 kgs per day.
7. One woman can sew 20–40 *dornams* (garlands) per day.
8. A group of 20 can uproot and pluck 1 acre of groundnut per day.

NOTES

1. Fieldwork was undertaken during 1983–4 and 1984–5 towards a doctoral degree from Syracuse University and was funded by the Ford Foundation. The views expressed are my own.
2. To spread water over as extensive an area as possible, at the inception of the canal the Irrigation Department authorized certain plots of land in each village, of the total number that could be irrigated, with a permanent right to water.
3. The Kammas are comparable to the Saswad Malis in western Maharashtra, described in Attwood (1984).
4. As the canal is a run-of-the-river system, the date of release is also dependent on whether rainfall in the upper catchment areas has been sufficient to swell the river.
5. There are a few rice mills but there has been little tractorization, introduction of irrigation pumps, etc.; in the limited areas where these and fertilizer/pesticide sprayers have been introduced, they have displaced women's labour.
6. Atchi Reddy (1989) notes that there are *maleta*s or female seasonal or annual servants in Nellore district of Andhra Pradesh, but that the institution is unique to Nellore.
7. According to the Kamma women, it is customary for landowners to feed agricultural labourers in the deltas, from where the Kammas immigrated. This is not the practice in the uplands where they now live, and they therefore have one less household task to perform.
8. This point has also been made by Kishwar (1984), but she emphasizes male alliances with the immediate and extended kin group.
9. Some researchers have contended that real incomes have decreased (Agarwal, 1988; Bardhan, 1981). Others (Lal, 1976; Chambers and Harriss, 1977) have shown that real incomes have increased. According to Government of India (1979, 1981) reports, all-India annual real earnings from all agricultural work for men and women declined over the period 1964–65 to 1974–75. Specifically for Andhra Pradesh, these reports show a decrease in the annual real earnings for men from 246.8 to 198.2 but an increase in the annual real earnings for women from 88.4 to 104.8. It is noteworthy that these government reports are for the whole state of Andhra Pradesh, not the specific research area, and that very often they are biased against women (Agarwal, 1985).

10. According to Gopalan (1981), milled rice contains 6.8 gms of protein, 0.6/100 gms minerals, 10 mg calcium, and 3.1/100 gms iron in comparison to 10.4 gms protein, 1.6/100 gms minerals, 25 mg calcium, and 5.8/100 gms of iron in jowar.

BIBLIOGRAPHY

Bina Agarwal, 'Water Resource Development and Rural Women', mimeo, New Delhi: The Ford Foundation, 1981.

———, 'Work Participation of Women in the Third World: Some Data and Conceptual Biases', *Economic and Political Weekly*, Review of Agriculture, 20(51 & 52): 21–8, December 1985.

———, 'Neither Sustenance Nor Sustainability: Agricultural Strategies, Ecological Degradation and Indian Women in Poverty', in Bina Agarwal ed., *Structures of Patriarchy: State, Community, and Household in Modernising Asia*, New Delhi: Kali for Women, 1988.

D. W. Attwood, 'Capital and the Transformation of Agrarian Class Systems: Sugar Production in India', in Desai, *et al.*, ed., *Agrarian Power and Agricultural Production in South Asia*, Berkeley: University of California Press, 1984.

———, 'Poverty, Inequality, and Economic Growth in Rural India', in D.W. Attwood, *et al.*, ed., *Power and Poverty: Development and Development Projects in the Third World*, Boulder: Westview, 1988.

Alan Beals, *Village Life in South India*, Chicago: Aldine, 1974.

Lourdes Beneria and Gita Sen, 'Accumulation, Reproduction, and Women's Role in Economic Development: Boserup Re-visited', *Signs*, 7( )2: 279–98, 1981.

Andre Beteille, *Studies in Agrarian Social Structure*, Delhi: Oxford University Press, 1974.

Ester Boserup, *Women's Role in Economic Development*, London: George Allen and Unwin, 1970.

Census of India, *Andhra Pradesh: Revised Figures of Population*, New Delhi: Government of India Press, 1984.

Robert Chambers, and John Harriss, 'Comparing Twelve South Indian Villages: In Search of Practical Theory', in B. H. Farmer, ed., *Green Revolution? Technology and Change in Rice Growing Areas of Tamil Nadu and Sri Lanka*, London: Macmillan, 1977.

B. D. Dhavan, *Irrigation in India's Agricultural Development: Productivity, Stability, Equity*, New Delhi: Sage, 1988.

Goran Djurfeldt, and Staffan Lindberg, *Behind Poverty: The Social Formation in a Tamil Village*, Scandinavian Institute of Asian Studies Monograph Series No. 22, London: Curzon Press, 1975.

T. Scarlett Epstein, *Economic Development and Social Change in South India*, Manchester: Manchester University Press, 1962.

Francine Frankel, *India's Green Revolution: Economic Gains and Political Costs*, Princeton University Press, 1971.

C. Gopalan, *Nutritive Values of Indian Foods*, Hyderabad: National Institute of Nutrition, 1981.

Kathleen Gough, 'Changing Agrarian Relations in Thanjavur, 1952–1976', *Kerala Sociological Review*, 1977.

——, 'Agricultural Labour in Thanjavur', in Joan Mencher, ed., *The Anthropology of Peasantry*, Bombay: Somaiya Publications, 1982.

Government of India, *Rural Labour Enquiry 1974–75, Final Report on Wages and Earnings*, Chandigarh: Ministry of Labour, 1979.

——, *Rural Labour Enquiry 1974–75, Final Report on Employment and Unemployment*, Chandigarh: Ministry of Labour, 1981.

John Harriss, *Capitalism and Peasant Farming: Agrarian Structure and Ideology in Northern Tamil Nadu*, Bombay: Oxford University Press, 1982.

Madhu Kishwar, 'Introduction' in M. Kishwar and R. Vanita, eds., *In Search of Answers*, London: Zed, 1984.

D. Lal, 'Agricultural Growth, Real Wages, and the Rural Poor in India', *Economic and Political Weekly*, Review of Agriculture, 47–61, June 1976A.

David Ludden, 'Economic Development and Social Change in Indian Agriculture', in D.W. Attwood, *et al.*, ed., *Power and Poverty, Development and Development Projects in the Third World*, Boulder: Westview, 1988.

John Mellor, *The New Economics of Growth: A Strategy for India and the Developing World*, Ithaca: Cornell University Press, 1976.

Joan Mencher and K. Saradamoni, 'Muddy Feet, Dirty Hands: Rice Production and Female Agricultural Labour', *Economic and Political Weekly*, Review of Agriculture, 17(52): 149–67, December 1982A.

Maria Mies, *Patriarchy and Accumulation on a World Scale: Women in the International Division of Labor*, London: Zed, 1986.

B. D. Miller, *The Endangered Sex: Neglect of Female Children in Rural North India*, Ithaca: Cornell University Press, 1981.

M. Atchi Reddy, 'Female Agricultural Labourers of Nellore, 1881–1981, in J. Krishnamurty, ed., *Women in Colonial India: Essays on Survival, Work and the State*, Delhi: Oxford University Press, 1989.

Gita Sen, 'Women Workers and the Green Revolution', in Lourdes Beneria, ed., *Women and Development*, Geneva: International Labour Office, 1982.

Burton Stein, *Peasant, State, and Society in Medieval South India*, Delhi: Oxford University Press, 1980.

7

Contradictions of Gender Inequality: Urban Class Formation in Contemporary Bangladesh[1]

SHELLEY FELDMAN

INTRODUCTION

It has been argued that gender inequality in the labour market, particularly in export processing zones, can be ascribed to the domestic division of labour and the recreation of domestic patriarchal relations in the work place. This argument is premised on the assertion that the articulation between reproductive activities and productive activities structure forms of gender inequality. This paper explores this assertion through an analysis of the social construction of gender relations in an emergent labour force in urban Bangladesh. Focused on the deployment of a new cadre of export workers in apparel manufacturing in Dhaka City, the paper examines the social and economic characteristics and family labour histories of this work force to identify the processes which have shaped gender differences among those employed in export manufacturing.

The paper also examines the transformation of the rural economy as this has provided the context within which male and female employment demand has increased among all segments of the rural population. Special attention is drawn to the relationship between rural and urban labour opportunities in illuminating the gender and class basis of contemporary production relations and to the importance of the ideology of *purdah* or female seclusion in shaping the configuration of the urban labour market. Weaving together an assessment of the changing configuration of the rural economy and recent industrialization initiatives, with an appreciation of the cultural construction of gender relations, the paper suggests potential contradictions posed by increasing the demand for female workers in the context of the growing importance of Islam in contemporary state practice.

The paper is divided into three sections. The first section employs Pearson's notion of the 'social construction of a labour force' to highlight the particular patterns of gender relations that characterize export production in Bangladesh. Of special importance in this context is the role of Islam and patterns of female exclusion which have shaped the export enclave gender division of labour. This approach differs markedly from one that asserts the needs of a capitalist enterprise and assumes that there is a ready labour force, a reserve army of labour, to fill the required posts created by an emerging industrial sector. Such an approach also differs from one that assumes that the gender division of household labour is simply reproduced by the workplace division of labour.

The second section examines the political economy of Bangladesh as this provides the context for analyses of economic restructuring and patterns of industrialization. This section summarizes information on the transformation of the rural economy, given the increased capitalization of agriculture and industry, as this has provided the context within which the demand for non-farm employment has increased. The last substantive section identifies the configuration of the industrial labour force and the nature of labour market dynamics in a stagnant industrial sector dependent on a competitive global economy for its reproduction.

THE SOCIAL CONSTRUCTION OF THE LABOUR FORCE

The availability of a labour force to fulfill the changing employment demands of a restructured global division of labour and thus various nation-based labour markets has been most notably discussed in the literature on the reserve army of labour. This available labour pool or 'marginal mass' is envisioned to move in and out of the labour market in correspondence with the changing patterns of capital accumulation (Bennholdt-Thomson, 1984). Theoretical investigations of the reserve army of labour, while helpful in exploring the dynamics of labour power as a commodity, have nonetheless failed to explore how this labour pool, composed of a particular gender, age and skill base, is created to correspond to structural changes in the economy, including new demands for female employment. Said another way, how can we begin to understand the ways in which a reserve army of labour is altered, shaped, and transformed by a changing political economy? In addition to the structural determinants which shape the labour supply and labour demand at a given point in history, how do normative expectations and behaviour proscriptions enable or constrain the mobility of this marginal mass? Pearson, in her insightful work on female workers, has argued that the movement of the reserve army of labour in and out of a changing national and international economy is a socially constructed process. In a critique of the new international division of labour and the incorporation of women into a global economy she poses the issue this way (1988: 450–451):

... the analysis of the new international division of labour has ignored the complexities and contradictions of producing the desired social relations of production involved in creating a new sector of waged labour . . . (Prior interpretations of labour force dynamics have not) acknowledged that either capital or the State might need to intervene to deliver the suitable labour required; (rather), it has been assumed that (an available labour force) was axiomatic on the existence of high levels of unemployment or underemployment in the Third World... (this debate also) assumed that the absence of industrial employment for women in the intermediate economic history of a country meant that there would be no problem in making... labour available in the quantities and qualities required.

An analysis of the relationship between the changing configuration of the Bangladesh labour force as this corresponds to changes in industrial production can benefit from a focus on the social construction of the labour force and urban class formation, since these concepts neither presume a determinant relation between these processes nor ignore the role played by state practice in shaping new patterns of hegemonic control. Using as our point of departure the social construction of the labour force, in other words, invites attention to the relationship between structural change and cultural practice. In this context we might ask: What are the normative constraints on women's employment imposed by Bengali cultural practice and Islamic interpretations of purdah and female seclusion? How hegemonic are these constraints on rural women's mobility and how have they been altered by new policies to spur investment and growth in the industrial sector and by the increased demand for low wage industrial workers? It is to an exploration of these issues that we now turn.

Characteristic of analyses of Bangladeshi women is the marked division between reproductive and productive labour. Included in the domain of reproductive activities are biological reproduction as well as the social reproduction of the relations of production and of society in general (Edholm *et al.*, 1977). In Bangladesh the normative assessment of female reproductive tasks has been centered on women's responsibility for cooking and crop processing for consumption, household maintenance responsibilities, and child care. This narrow definition of domestic work underestimates the complex set of activities embraced by the central role of domestic labour and the organization of the family in shaping the social organization of people and institutions in subsistence production. A narrow definition of domestic work within a subsistence production also ignores women's contribution to crop production and crop processing in Bangladesh since their engagement in market activities and exchange has been mediated by male household members.

More important for the present discussion than the forms of female participation in home-based subsistence production is the fact that the spatial distinction between the male and female domains of production has lent credence to the normative assessment of women's work and women's access to the labour market. For example, the seclusion of women from public exchanges has been reinforced by the location of reproductive and subsistence activities which, until quite recently, have been the private and invisible domain of the household where women have been constrained in the ways in which they could engage in remunerated activity and were assumed to be guarded in their interactions with non-familial males. Women's circumscribed behaviour has been assumed to begin at puberty and extend to widowhood. Productive tasks and remunerated labour exchanges, on the other hand, have been viewed as part of the public domain and include what are conventionally categorized as employment activities. This has traditionally been the domain of men and includes their control of political power and decision-making. The distinction between reproductive and remunerated labour exchanges, while heuristically useful in analyses of gender inequalities, masks the productive contribution women make within the household and the ways women's productive obligations, often unaccounted for in discussions of the household division of labour, also shape women's participation in the public domain.

Counterposing either reproductive and productive activity or private and public work domains tends to treat these arenas of production as non-reflexive or autonomous. An alternative to this aprocessual and reified view of the gender division of labour assumes that gender relations are shaped by the conflicting and at times complementary demands of different forms of production. This suggests that gender inequality and the particular configuration of the social relations they engender can neither be 'read off' from the economic base nor can they be understood ahistorically. Gender relations, in other words, do not simply reflect an ideal-typical view of a subsistence-based division of labour nor do they merely express a specific set of traditional or feudal relations. Rather, such relations are historically emergent in the context of a diverse range of existing social practices. As such, these practices are constructed in the context of prior expressions and relations of inequality, the organization of new production structures, and the ideological complex that serves to elaborate and legitimate new patterns of social interaction.

THE RURAL ECONOMY

Bangladesh is a country dependent on agricultural production to meet its financial and trade requirements and to provide approximately 80 per cent of total employment. Jute and other agricultural resources, including fisheries, have contributed a major proportion of the country's purchasing power and account for the majority of its trade. However, declines in the demand for agricultural labour and the stagnation of jute and leather exports, coupled with only modest increases in tea exports, suggest an overall decline in agricultural labour demand (de Vylder, 1982: 29). Such declines are likely to increase the demand for off-farm and non-farm[2] rural employment among a growing segment of the rural population.

The significant growth in agricultural underemployment and unemployment is indicated by changing landholding patterns. By 1978, the total number of landless households was estimated at approximately 29 per cent of the rural population. If we add to this category households categorized as functionally landless or who own less than one acre, the proportion of the rural population likely to require off-farm employment increases to almost 80 per cent. Those owning between one and 2.5 acres are also likely to increase their demand for non-farm work since it is assumed that 2.5 acres is the minimum required for economic viability. In other words, among households with declining land holdings or declining agricultural incomes there is likely to be an increasing demand for non-farm employment to subsidize subsistence and family reproduction (Jannuzi and Peach, 1979; UNDP, 1988).

Almost 67 per cent of non-farm rural households secure employment as agricultural labourers. It is estimated that 41 per cent of small farm households and even a small proportion of farmers from households with more than 2.5 acres supplement agricultural incomes with agricultural wage labour (UNDP, 1988: II–3). However, since the demand for agricultural labour is not anticipated to increase by more than between 200,000 and 300,000 person-years annually, the number of households likely to secure work as agricultural labourers may decline and thus further increase the demand on the urban and rural non-farm sectors (UNDP, 1988).

Not only have males from under-subsistence farm households sought employment as agricultural labourers in addition to employment as construction workers and daily labourers in unskilled rural non-farm work, but poor women are increasingly represented among those seeking employment. In one study, for example, working women were found to represent between 18 and 25 per cent of the total number of households in representative villages from four districts in the country (McCarthy, 1980).[3] This finding is corroborated by a recent survey of 3,965 households from 20 districts, where it was found that 12 per cent of the households surveyed had women engaged as agricultural labourers with an additional 12 per cent engaged in non-farm activities (UNDP, 1988: II–5–6). The numbers of women employed in the rural areas, however, vary significantly by size of agricultural holding. For instance, this same study revealed that among landless households 60 per cent had women engaged in agricultural or non-farm employment (UNDP, 1988). Food for Work studies of very poor women also reveal that they are among a growing number of the rural poor demanding and securing employment on work gangs and maintenance crews (Chen and Ghaznavi, 1977; Marum, 1982; CARE, 1983).

The high proportion of women seeking employment is quite unusual for Muslim Bangladesh where rural women's work has been traditionally 'hidden in the household' and tied to agricultural processing and production. In the 1970s women did not secure work in field cultivation although by the early 1980s small pockets of poor women were visible in the harvesting of chillies and potatoes in Comilla and Chittagong. More recently still, women have been able to participate in the marketing of produce through the creation of special facilities for them. For a small minority of educated women, work has traditionally been restricted to employment in the education, health and welfare sectors where they remained at home until marriage and worked in their own or an adjacent community. These opportunities have been few and limited to a minority of educated women from middle peasant families who have been able to educate their daughters and help secure employment for them. It is only recently that opportunities for non-farm employment have been recognized as a vehicle of status construction, but more on this latter point later.

The increasing visibility of women in agricultural production and in professional government services suggests the interest among women of all classes to secure wage income. For women from landless households, any opportunity for employment is taken and among the very poor, rural employment opportunities have generally been limited to work in Food for Work programmes or as agricultural labourers. These latter opportunities are both seasonal and low paid and do not carry the possibility of permanent employment. Among the lower middle classes efforts to secure work generally parallel the erosion of economic and personal security that has characterized Bangladeshi life since Independence.

These emergent trends in the pattern of female labour force

participation do not represent a cataclysmic ideological challenge to traditional Islamic practices following the introduction by the government of a New Industrial Policy (NIP) in 1982 which included the establishment of export processing enclaves. Rather, the growing demand for non-farm employment is one indication of the erosion of material and social resources and networks for a growing proportion of the rural population.[4] The term erosion is useful because it suggests a slow but continuous process of increasing impoverishment. It suggests, for example, that although census data have not reported the extent and type of female employment shifts, women have been increasingly involved in various work relations including but not limited to field cultivation for wages. What studies of female employment are available tend to focus on the poor whose demand for work is mostly realized by their increased participation in field agriculture, as household labourers, and as family-based workers in cottage industries. Middle-class rural women have been increasingly represented among government employees in the health and education sectors. Very little attention has been focused on women in the private sector, an arena only recently expanding.

An emphasis on the slowness of the process of change in the agrarian structure and its attendant social relations and of the steady expansion of women's participation in the labour force differs from analyses which posit cataclysmic changes in women's position in Bangladesh (Abdullah and Zeidenstein, 1981; Wallace, 1987). These latter studies argue that it is only very recently that women have sought and been allowed to engage in employment generating activities, and view women's employment as a consequence of epiphenomenal events such as the independence war of 1971, the famine of 1974, or the recent floods rather than as a process of the changing agricultural and non-farm production environment. While particular events may serve as catalysts for increasing women's labour force participation rates, women's responses to these events have been shaped by the capitalization of agricultural production, the increasing proletarianization of subsistence production, and the declining family wage.

The structural shifts in the rural economy highlight concomitant redefinitions in forms of acceptable female behaviour. For example, it is only recently that female employment has come to be seen as a negotiable attribute in marriage arrangements. Traditionally, middle strata subsistence producers depended on a daughter's labour in maintaining agricultural production, and the family accrued status by keeping women out of the labour market. Such families were seen to have the resources sufficient to secure a daughter's marriage to a family whose son could either earn or generate sufficient agricultural income so as not to depend on the remunerated labour of his wife.

In the context of the transformation of the domestic production structure and declines in the number of subsistence and surplus producers, an increasing dependence on wage income among a growing proportion of rural households, and declines in the value of the wage, the 'family wage' is now insufficient to realize the costs of reproduction. This results in an increased dependence on more than one family member being required to secure wage employment to meet subsistence needs and thus increases women's demand for work (Feldman, 1989). For middle peasant, middle-class rural families, insufficient rural non-farm professional employment has forced increasing numbers of their members to consider urban employment since there has been limited growth of both the private and public rural manufacturing sectors. As we will see below, the commitment to export-led industrialization under the NIP has created a demand for female workers in Dhaka where once women were hardly visible among the urban labour force.

Before turning to the next section, however, it is important to note that to hypothesize a relationship between changing cultural practices and changing forms of economic security implicates the reflexive nature of this relationship. An examination of this relationship as a dynamic and contradictory one also focuses attention on the ways in which patterns of global economic restructuring are represented by specific nation-based processes of capitalist development which are mediated by historically specific cultural practices. This relationship, which is expressed in Bangladesh as the contradiction between traditional forms of female subordination and female seclusion and the newly created demand for female workers, challenges what Pearson has referred to as the axiomatic principle that there is a ready and available labour force to correspond to the mobility of capital. This contradictory relationship also challenges those who argue that there is a logic to capital accumulation which is structurally determinant in so far as it includes a reserve army of labour and appropriately trained labourers available to meet the changing demands of capitalist accumulation or that capitalist development processes include 'occupational slots "preordained" as it were, for the female gender' (Bennholdt-Thomson, 1984; Hossain et al., 1988). These arguments assert, but do not explain, how this reserve army is created and maintained so as to be available, in appropriate proportions and with appropriate skills, on demand. In this context, the flexibility of ideological hegemony and the social construction of gender are assumed rather than explained.

What makes an examination of the contradictory relationship between economic restructuring and patterns of female seclusion especially salient in the recent period in Bangladesh is the importance the Ershad regime gave to using Islam as a mechanism for maintaining, and maybe strengthening, forms of gender exploitation while 'freeing' women to engage in modern sector activities. At the base of this contradiction also lie new definitions of female status and marriageability, new forms of mobility, and gendered patterns of class formation and class relations.

THE INDUSTRIAL ENVIRONMENT

Efforts to expand rural employment opportunities began in the 1960s with the establishment of industrial estates in the major cities and provincial towns. Support for these industrial estates was intended to provide a social and technical infrastructure to support medium and small scale enterprises in rural towns, to provide investment opportunities for local entrepreneurs, and to generate employment for a growing rural population. Twenty years later it was found that these industrial estates were never made fully operational: they had only partially been supplied with water, power lines, and roads, and credit facilities were insufficient to provide the incentive needed to generate capital investment and employment. Moreover, except for the Rajshahi indus- trial estate, where women have been employed in silk reeling, spinning and weaving, women have been unable to find regular employment on the estates (Feldman, 1984).

During the 1970s other national policy initiatives to stimulate industrial growth also met with only partial success. For expanding female employment opportunities the Bangladesh Small and Cottage Industries Corporation (BSCIC) began to work with women to increase the type of handicrafts and the quality of the goods they produced. Karika, a women's marketing co-operative, began in the mid-1970s to sell women's crafts to urban middle-class families in the capital city. In the latter 1970s a number of domestic and foreign projects expanded the supply of traditional crafts that engaged women by opening small boutiques and shops for sale to those in the larger cities and towns. Eventually, these crafts, often produced by poor and lower middle-class rural women either privately or via middlemen through a putting out system, were sold through private shops in the cities and towns of Bangladesh.

More recently still, semi-autonomous and non-governmental organizations have encouraged expanded production among village and provincial town entrepreneurs to increase employment and generate a demand for hired labour (Feldman and McCarthy, 1984). Among the most notable of these efforts has been the Grameen Bank Project which has successfully provided credit to poor rural women (Grameen Bank, 1985; Hossain, 1985). While these efforts have successfully generated a self-employed cadre, they have not significantly increased the demand for non-farm labour. Medium size industries have also been supported by government and foreign donors to mobilize local resources for productive activity and discourage capital flight. Women have not been targeted for this entrepreneurship programme and thus have been excluded from access to subsidized credit and technical support (Feldman, 1989). However, to the extent that employment has been generated by these firms, female workers have found work in them.

These changes in domestic policy and resource allocation to the non-farm sector were shaped during the period of significant agrarian change characterized by increased landlessness and a changing agricultural labour market. Concommitant with these changes were those stimulated by significant global economic restructuring and the redivision of the international labour market. Economic restructuring is represented in the literature in the arguments about the new international division of labour (Frobel *et al.*, 1980), export processing enclaves and the rise of Third World manufacturing, and the global assembly line (Elson and Pearson, 1981; Nash, 1983; Fuentes and Ehrenreich, 1984). The fragmentation of the labour process in manufacturing, the quotas placed on the newly industrializing countries of Hong Kong, Singapore, South Korea and Taiwan (NICs) by the United States, and the search for low wage workers have been salient themes in the promotion of policies adapted by Bangladesh to generate and support an export-led growth development strategy. It also corresponded to the promotion of export manufacturing, the limited success of rural industrialization initiatives, and pressure from the IMF and the World Bank to increase foreign exchange earnings and to redirect the development strategy away from one focused on import substitution.

The antecedents of the policy shift from import substitution and the expansion of agricultural capacity to export-led industrialization and privatization of industrial production can be rooted in the decline of Sheik Mujib (1971–75). Under the regimes of both Zia Ur-Rahman (1975–81) and General Ershad (1982–91) the opportunity for Bangladesh to be integrated within the orbit of global manufacturing served to generate increased financial support for urban infrastructure, technical and financial assistance for industrial reform, and to provide a mechanism to support urban elite interests through the provision of industrial credit, tax holidays, and import licenses. This shift assumes that such policy reform will enable Bangladesh to repay its loan commitments and meet the demands of structural adjustment lending as well as realize the promise of export promotion to expand employment opportunities.

Bangladesh could only be incorporated into the global manufacturing network with significant restructuring of the industrial sector, a national commitment to tax, credit and trade reform, a recognition that urban migration and attendant levels of urban poverty would likely increase, and a willingness to integrate women into the industrial labour force. The NIP of 1982 integrated these efforts to stimulate industrial development and to generate what has been assumed to be a more productive national economic base. Another important element in this export promotion strategy has been the assumption that Bangladesh can best meet the needs of the rural disenfranchised because of a comparative advantage in the availability of low skill wage labour (Gray, 1990). The magnitude of people available to be drawn in the low-wage manufacturing coupled with the pressure imposed by the state to limit the labour union movement and to ignore minimum wage standards, as well as the lack of quotas imposed on other countries producing manufactured goods for export, have provided the structural preconditions for Bangladesh to be directly integrated in the 'global assembly line' (Fuentes and Ehrenreich, 1984). A recent article emphasizes the continued advantages of the cheap labour and limited enforcement of wage and labour standards in the country (Gray, 1990).[5]

A PROFILE OF INDUSTRIAL WORKERS IN THE DHAKA EPZ

Studies of employment in export processing zones (EPZ) reveal that forms of gender inequality include differences in income, in the types of contracts men and women enjoy, in the rank and type of service open to men and women, and in the nature of the labour relations in which men and women are engaged (Edgren, 1982). Such studies also indicate that employment in export production has contradictory effects for women. On the one hand, such employment 'frees' women to compete and labour on the market and thus 'liberates' them from their dependence on unvalued or undervalued subsistence production. On the other hand, EPZ employment is characterized by a gender segmented labour market where women are limited to low wage work, limited job security and constraints to upward mobility (Elson and Pearson, 1981). Such inequality is also depicted by often unfavourable working conditions which include overcrowding, poor ventilation, and limited rest-room facilities as well as unfavourable contractual terms such as few or limited work breaks, very short lunch hours, and undefined holiday and vacation benefits. Moreover, given the nature of EPZ employment, often on short term and quick turnover contracts, female production workers are generally required to work overtime to meet deadlines. This could extend the work day by as much as 12 hours (Feldman, 1984; Phelan, 1986).

Comparative information on EPZ workers reveals that the workforce can be described as comprised of young, educated, unmarried women (Heyzer, 1988). In Bangladesh the labour force is similarly characterized. However, what is particularly interesting in the Bangladesh case is that unmarried women, especially those women who come from educated, rural, middle-class families, have generally been kept off the labour market by their parents since family status was established and maintained by a father's ability to keep his daughters off the labour market. This expression of status construction poses distinct problems for women's employment in export processing enclaves and for their future marriage opportunities since young single women's employment contradicts those established female behavioural proscriptions and practices which have traditionally kept women from access to urban employment.

The notable absence of women in skilled and formal sector employment in the past therefore suggests that the recomposition of the labour force will be articulated in quite specific ways within the structural context of Bengali purdah. Moreover, women's relative invisibility in subsistence labour, petty commodity production and the informal sector, at least until recently, suggests the degree to which their visibility in export manufacturing represents a new expression of their participation in the production process and a new expression of female propriety. Employment in the EPZ, for example, is associated with women who come to the urban areas without their families. It is generally first time employment for these young women who, in the context of securing such work, often face male harassment and new demands on their personal and social relations. This includes an expectation that they speak freely with men as well as socialize with them.

More distinctive is the fact that new marriage expectations have been generated by a woman's control of work and income. For example, where a woman's status was defined solely by her family resource base and her family's control of her access to the labour market, the security of modern sector work and income has come to be seen as important in negotiating marriage arrangements. Female employment also has begun to include new expectations of marriage partners and of domestic and work obligations for women once they are married. Additionally, the opportunity to socialize with members of the opposite sex has led to a decline in arranged marriages and an increase in love marriages. This poses a distinct challenge to families who have been able to use the marriage of children to generate and solidify class and kinship ties. In short, the articulation of these new behavioural practices and changing expectations generates contradictory effects for women and their families.

Estimates of the total number of employees in the export garments sector range between 168,000 and 210,000, 80 to 90 per cent of whom are women (Banu, 1986; Phelan, 1986). Of the total sample,[6] the majority are unmarried women, 53 per cent of whom were between the ages of 10 and 19 years. An additional 29 per cent were between 19 and 25 years. Men between 10 and 19 accounted for 21 per cent of male workers and an additional 48 per cent bring the total of males under 25 to 69 per cent. Men too tend to be unmarried and work in export manufacturing as their first employment. As suggested above, the age and marital status profile of those engaged in Bangladesh's export sector parallel worker characteristics found in other EPZs (Elson and Pearson, 1981; Nash, 1983; Fuentes and Ehrenreich, 1984), but unlike workers in other countries the Bangladeshi female labour force represents lower middle and middle rural educated women who recognize that while wages are low, they are assumed to increase after an initial apprenticeship period and eventually lead to permanent employment.

Before turning to a discussion of the interests and relations represented among the new industrial labour force, it is important to highlight the ways in which women's participation in the Dhaka export processing zone represents constructions of new normative standards of behaviour which build upon and reflect shifts in the accoutrements of and discourse about purdah and Islam (Feldman and McCarthy, 1983). For example, one way in which appropriate female Islamic behaviour has been supported in the context of female employment is to have women wear the *burkha*, or veil, on trips between home and work and, as 'appropriate' when meeting with male colleagues. In this context, the burkha has become the vehicle of mobility, almost literally, as it enables women to be transported in an appropriate style while also enabling them to seek and secure income. How have new models of appropriate behaviour characterized the ways in which women seek and secure work in the export enclave?

One way this is enabled is reflected in the recruitment

strategies of new entrepreneurs which maintain patriarchal values while challenging the forms in which these values are expressed. Access to employment takes various forms; newspaper advertisements, word of mouth or kin and village relations with the factory owner or *malek*. Among established manufacturers, using kin and village relations to generate a labour pool is the most common means of hiring labour. This is accomplished by the concentration of workers from particular districts in selected factories. Workers from Kushtia District, for instance, comprise the majority of the labour force in one of the country's largest garment factories owned by a native of Kushtia District.[7]

The parents of many women who are recruited directly from the home village or district of the employer are assured by the malek that he will serve as guardian for their daughters. His assurance also includes that safe and proper housing will be available to the new employee. Such employers also suggest that young women will live near the factory, with women they know, and will work with other women on the factory floor. Since employers generally represent 'village heroes' who have found success and therefore constructed a reputation of good standing in their home areas, especially if they offer villagers opportunities for urban employment, the valuations that they hold are generally not questioned. Thus, a cadre of female workers are recruited for industrial production within the context of a changing but not radically altered interpretation of female propriety under male guardianship and patriarchal control.

Not surprisingly, this recruitment strategy builds upon the relatively recent development of a business community in Dhaka City where manufacturing firms are generally owned and operated by first or second generation urban families still identified with and tied to the countryside. These generational ties to the countryside are often reinforced by the investments new urban entrepreneurs make in their home village. These investments may include the building of a *madrassa* or mosque which help reinforce the good standing of the entrepreneur in the village and establish their credibility as good Muslims sincerely committed to the well-being and honour of village kin.[8] For workers these connections with respectable villagers represent the hopes and possibilities of upward mobility and urban sophistication. These contacts also provide a vehicle for worker mobility and the creation of new networks of control for them.

The benefits which accrue to the malek because of his control of urban resources such as employment and housing facilities include worker obligation and commitment since the malek is assumed to have done a worker's family a favour by hiring one of their children or relatives. In exchange for this employment opportunity the worker is often pressured to favour management over labour in decisions about wages, overtime, health benefits and work conditions. To maintain work commitment, the factory owner may also offer bonuses and extra holidays to village relations in exchange for their willingness to serve as information conduits between other labourers, supervisors and owners. This strategy of obligation and control has proven to be particularly important during times of strife and worker disaffection as it enables owners to better control their work force and helps to remove or fire those involved in union activity or efforts to build worker solidarity.

For workers, a job in export manufacturing provides an opportunity to leave the countryside with guaranteed employment where wages, relative to opportunities in the rural areas, are presumed to be higher. Since opportunities for urban employment are still relatively few, women often gratefully accept EPZ employment without recognizing the costs involved in the terms established by the malek. For one, first generation urban dwellers are unprepared for the costs of urban life, with food, housing and transport being costlier in the city than in the rural villages and towns. Additionally, there is an absence of urban housing, especially for women, as hostels are few and landlords are hesitant to rent to unmarried women. In this context there is a sense of physical and moral insecurity for the recent urban migrant unmarried women which tends to reinforce the family's reliance on the judgement and beneficence of a malek to assure them of their daughter's safety. Such relations tend to give the malek even greater control of the workforce than would likely be the case in a formal contractual arrangement. Moreover, the informal contracts established between employers and workers often maintain women as apprentices for longer than is generally expected as such wages are kept well below subsistence for sometimes well over a year. A 1990 report indicates that wages in the garment industry are as low as US $ 0.09 per hour (Gray, 1990: 15).

Since export sector work is a new opportunity for almost all workers, and many are first generation urban residents, labourers often maintain social networks with rural kin as a source of social support and solidarity. The dependencies which develop in these social networks, especially among those employed in the same factory, often parallel and reinforce many of the social relations patterning village inequalities (McGee, 1982). These dependencies often serve to extend and reinforce forms of social control either by an urban elite over an emergent industrial working class, especially those representing village kin, or by other workers who claim a degree of obligation from those for whom they have been an important tie. For women, these relations of dependence are contradictory: on the one hand, they enable forms of solidarity building among women as workers and as new urban dwellers, including new bonds of friendship which had little opportunity to flourish in the context of rural familial control and obligation. On the other hand, these ties and patterns of dependence maintain previous expectations of appropriate female behaviour and enable mechanisms of direct social control by male workers and supervisors.

These relations of dependence represent dynamic processes of change. As Tilly (1986) has recently noted: 'we ought to think of migration as we think of community structure, not simply reducible to individual characteristics and intentions'. And, as Tilly goes on to suggest, even when people migrate alone they usually draw on and use the information from network members that have migrated before them. The networks that are thus generated are likely to reinterpret relations of gender inequality rather than challenge the bases of this inequality. This is especially likely when women represent those workers who have limited chances for mobility, earn low wages and are least organized to protect their own interests.

Given the interdependence of village and urban kin, and the normative proscriptions that characterize these labour relations, who are the beneficiaries of these urban connections? What rural interests and relations are represented among the new labour force? One surprising finding in this study is that export workers do not represent the poorest of rural families who may seek employment as a last resort. Of our sample population, 79 per cent come from families owning some landed property; 53 per cent, however, own less than 2.5 acres and are thus likely to be involved in diversified household income earning activities. The amount of land owned by export workers compared with the distribution for the nation as a whole is summarized below (Table 1).

The most striking feature in this table is the pattern of gender differences in land ownership. Among female workers in the sample, 45 per cent come from families without land and for those whose families own land, most represent small and marginal agricultural producing households. For men a comparable figure is only 9 per cent. Thirty per cent of the male workers come from surplus producing households while less than 7 per cent of the female workers come from such households. Of those with land, landholdings range from homestead plots, some with gardens, to the ownership of cultivable acreage with an orchard or pond. A small number of households rent or sharecrop land in and out. The majority of workers from large landowning households have family members who are directly engaged in agricultural production either as producers or managers. In other words, female workers were more likely than men to come from landless families, and of those with land, women's families tend to own less land than their male counterparts.

TABLE 1
NATIONAL AND SAMPLE COMPARISON OF LAND OWNERSHIP
OTHER THAN HOMESTEAD LAND

Number of acres	National Per cent of total	Sample population Per cent of workers Males	Females
Zero	33	9	45
< 1	29	23	20
1–2	16	13	15
2–3	8	10	7
3–4	5	9	5
4–5	3	7	2
5–6	2	5	1
6–7	1	6	1
7–10	2	9	3
> 10	1	9	1

SOURCE: Adapted from Jannuzi and Peach, 1982 (Table D2).

Of women workers whose families own land, the majority represent marginal and subsistence households, families who traditionally were unlikely to encourage their unmarried daughters to seek employment outside the household compound. Among women workers whose families do not own land, it should not be assumed that they represent the poorest segments of the rural population. Twelve per cent of the male workers and 17 per cent of the female workers have fathers who are businessmen. Thus, they represent households which have already diversified their income earning strategies and thus are likely to exemplify petty bourgeois, urban interests. The proportion of Bangladeshi industrial workers representing non-agricultural rural families contrasts quite markedly with those of other Asian countries where women have been assumed to be from families still predominantly engaged in agricultural production (Fuentes and Ehrenreich, 1984).

For those families still tied to subsistence production, one reason for the rapid acceptance of urban employment by middle strata rural households is the limited opportunities to expand agricultural production. Families with land often diversify their income sources by encouraging male children to accept wage employment in the non-farm sector, including manufacturing, rather than fragment already quite small holdings. This is most starkly suggested by the relatively high proportion of male workers who come from landed households. Since men rather than women are likely to be quite mobile within the manufacturing hierarchy, employment in the newly industrializing sector has become a strategy for household income diversification for sons who previously would have joined government or military service but for whom these options were neither lucrative enough, accessible, nor of sufficient status to draw their commitment. Since more than 85 per cent of the males in the sample have some secondary education and a small proportion have college degrees, it is likely that sons have been encouraged to seek employment in export processing in order to diversify urban contacts, decrease dependence on the inheritance of landed property, and thus enable a family's land holdings to be passed on to a single son.

The shift from agricultural to non-farm occupations is indicated by a comparison between grandfather's and father's occupation. For women, all their grandparents were engaged in agricultural production but the majority of their fathers had already diversified their income securing activities and were engaged in non-agricultural, professional and semi-professional occupations. It is therefore not surprising that such families have educated their daughters since it is among marginal subsistence producers that female education has yet to become a normative goal. One explanation for this difference is that non-agricultural rural dwellers appear to be more mobile than their agricultural counterparts and thus see daughter's education as a vehicle for negotiating a marriage arrangement with a similarly endowed rural household and thus contribute to family status. Previously, educating a daughter did not necessarily enable her to seek and secure employment. Among marginal agriculturally producing households it was more likely that families would forge a marriage alliance with another landowning family if their family status could be represented by their ability to keep their daughters out of the labour market and only minimally educated.

From the point of view of Bangladeshi entrepreneurs, they seek to hire an educated work force despite the fragmentation of the labour process and the low wages characteristic of such employment. This follows the tradition of export processing workers in other parts of Asia and Latin America. Since the labour supply in Bangladesh far outweighs the demand for workers, hiring favours an educated middle stratum of the rural population and thus is accessible to only a minority of the rural population seeking non-farm employment. Presented below are national educational trends against which the sample population can be compared.

TABLE 2
COMPARISON OF NATIONAL AND SAMPLE POPULATION
DISTRIBUTION OF EDUCATIONAL ATTAINMENT LEVELS

	\multicolumn{8}{c}{Highest level attained}
	No schooling		Incomplete first level		Incomplete second level		Post secondary	
	(Natl)	(Smpl)	(Natl)	(Smpl)	(Natl)	(Smpl)	(Natl)	(Smpl)
Both sexes	82.3	5.9	10.0	20.4	6.9	65.0	0.9	8.7
Males	72.3	2.0	13.9	10.5	12.1	74.6	1.7	12.9
Females	93.0	10.8	5.4	32.8	1.4	52.9	0.1	3.4

SOURCE: Abstracted from UNESCO, *Statistical Yearbook*, 1978–79 (p. 45) and from survey data.

National figures reveal the high illiteracy rates for men and women: 72.3 per cent for males and 93 per cent for females. This contrasts quite starkly with our sample distribution of 2.0 per cent and 10.8 per cent, respectively, or a total of only 5.9 per cent of those employed in export processing being either illiterate or having attended only a madrassa (religious training). The majority, almost 65 per cent, have completed secondary school and an additional 8.7 per cent have attended higher secondary school or have a BA or MA degree. What is so startling about the numbers of educated women employed in the enclave, 89.1 per cent, is the limited skills required for work in the sector and the rapidity with which such employment has come to be seen as a vehicle for income security and a potential mechanism of mobility among this rural stratum, especially for rural women.

Also of interest is that 40.6 per cent of those employed in export manufacturing are either the eldest or second child in their families: 36 per cent of all males and 43 per cent of all females employed in export manufacture. These figures suggest the pressure on rural producers to diversify income sources to establish income security. Also suggested by these findings is that many middle peasant and rural petty bourgeois families are unable to earn enough in agriculture to support their male children, their wives and offspring, or to negotiate a good marriage for their daughters. Strategies for income diversification, however, differ by the resource base of the rural family, the gender of the worker, and the kind of connection the family has with a factory owner. For landless males, work in manufacturing is seen as a potential source of upward mobility especially if it has been encouraged by a village entrepreneur recruiting employees. For families with land, and with educated sons, such employment may be a means to diversify family connections in the urban, modern sector. For women mechanisms for intergenerational resource transfer have begun to include modern sector employment which helps to realize new household income diversification strategies and new opportunities for individual income security.

In sum, the landowning, non-farm employment, and educational patterns characterizing the export manufacturing work force suggest that male and female workers represent different segments of the rural population, are likely to maintain different linkages to their village home once employed, and are likely to face different forms of structured inequality in the labour force. Moreover, the mechanisms rural families employ to secure urban work, in response to changing opportunities in agriculture, highlight the complex relations involved in understanding processes of proletarianization and the relationship between rural social hierarchies and urban class formation. In other words, occupational diversification, once available to rural surplus producers who could hire labour and educate their children, has now become a strategy of middle peasant producers if they are to survive in agriculture and in non-agricultural rural productive activity.

CONCLUSION

How do we interpret these demographic patterns of the export manufacturing work force and processes of income diversification in the context of contradictory forms of gender inequality and urban class formation? New patterns of income diversification and income security reflect the transformation of agricultural production which have generated an insufficient demand for rural off-farm and non-farm workers and thus has increased the demand among rural dwellers for urban employment. The demand for work among women corresponds to declines in the family wage and the increasing need for individuals to take responsibility for the costs of their own reproduction. This increasing need for individual income security poses distinct problems for Bangladeshi women where the cultural proscriptions of purdah and patterns of subsistence production dependent on female labour within the household had previously generated the conditions for women's exclusion from the labour market.

In other words, with changes in the conditions of agricultural production, and new demands on women to secure wage income, the earlier proscriptions of Islam are insufficient to legitimate new patterns of women's work relations and income earning strategies. Thus, within the broad outlines of female seclusion and propriety, new expressions of appropriate female behaviour are emerging. While it is premature to confirm the configuration of the pattern that is emerging, the reconstitution of female behavioural norms is building on earlier patterns of female subordination and gender inequality which are being deployed to legitimate and maintain relations of inequality in the labour market.

While the growing demand for work has come to characterize the needs of a majority of the rural population, the ways in which female employment is secured and legitimated differs by the class and status of rural households. For example, for middle class educated women the ideology imbedded in modern sector employment—factory work among other women of similar status—builds on the cultural practice of female seclusion. The new relations of gender inequality which are emerging in this context are reinforced by the interdependencies of urban and rural networks and changes in intergenerational resource flows. The reconstitution of patterns of the control and subordination of women are also established by the ties which bind rural households to new urban entrepreneurs who use their control of the urban labour market to recruit workers and obligate them in ways which limit their independence and authority.

Employers recruit young women within the framework of their special access to housing markets and their willingness to serve as guardian for young female village kin. These conditions of employment build on the inexperience of rural women in the urban labour market and help to shape new patterns of gender inequality. This inexperience tends to create the conditions under which women come to be docile and disciplined workers, willing to accept tedious and repetitive work in exchange for the opportunity to secure work and generate conditions for their greater autonomy.

Such characteristics of the female labour force are neither biologically grounded nor simply passed on generationally. Rather, these characteristics are the product of patterns of socialization and differential access to resources which shape who can secure new employment opportunities and how this pattern of socialization is gendered in the relations deployed to construct the new industrial working class.

NOTES

1. Research for this paper is part of a larger project supported under Fulbright Grant Number 83–006–IC. Data was collected between March and November 1984 and supplemented in 1988. Special thanks are due to the College of Agriculture and Life Sciences, Cornell University, for support to code and analyse the data and to Alice Clark and Florence McCarthy for helpful suggestions and refinements of the arguments made.
2. The terms off-farm and non-farm refer to sharecropping and renting arangements and non-agricultural employment respectively. Unless specified, the terms will be used interchangeably.
3. Using the household rather than women as the unit of analysis helps to highlight the declining family wage and the growing dependence of households on more than one income to maintain social reproduction.
4. It should be acknowledged that some women have benefited from the changing agrarian production structure and the opening up of the labour market to educated women during this period. The point here is that for the growing majority, the erosion of economic and social supports changed the nature of their exploitation and the context of patriarchal control and authority.
5. The NIP encourages foreign investment through the establishment of an export enclave to attract both 100 per cent foreign operated, joint venture or 100 per cent domestic ventures in export manufacturing. As of 1984, 177 garment factories have been approved, 79 were awaiting approval and an additional 167 had submitted applications seeking approval for export production. Only six garment factories were joint venture operations, with partners from Singapore, South Korea, Hong Kong and India, one joint venture electronics firm with a partner from Sweden, and a leather factory that

was a subcontracting firm for the multinational Bata Shoes (New Nation, 1984; Feldman, 1984).

6. Unstructured, intensive interviews were held with workers and employers during nine months of 1984. Information was also gathered from members of the Planning Commission, the Ministries of Labour, Industries and Social Welfare, and the Export Processing Zone Board to provide a picture of the policy environment of which export workers are a part. Additional information from the international donor community who have financed and have provided technical assistance to help realize a new industrialization strategy for the country has also been collected to help outline the relationship between global restructuring and its particular expression in the political economy of Bangladesh since Independence. This research was made possible with the help of Fazila Banu who assisted the efforts throughout, and to six interviewers who spent till the wee hours of the evening interviewing workers in their homes.

7. A bias of the referred sample design is the likely concentration of respondents from similar factories or villages. This should be taken into account in analyses of sample distribution although the point about population distribution is confirmed by both owners and workers in the sector.

8. The point here is not to argue for a premeditated set of actions to win over, or create credibility among rural villagers, but to indicate the complex and contradictory processes involved in the creation of social legitimacy.

BIBLIOGRAPHY

T. Abdullah and S. Zeidenstein, *Village Women of Bangladesh: Prospects for Change*, Oxford: Pergamon Press, 1981.

Fazila Banu (Lily), 'Intensive Rural Works Programme-Bangladesh: Project Documentation of Agriculture Extension Programmes', DANIDA, NORAD, SIDA (mimeo), 1986.

BBS (Bangladesh Bureau of Statistics), 'Preliminary Report on Labour Force Survey, 1983–1984', Dhaka: Ministry of Planning, 1984.

Veronika Bennholdt-Thomson, 'Subsistence Production and Extended Reproduction', in Kate Young, Carol Wolkowitz, and Roslyn McCullagh, ed., *Of Marriage and the Market*, Boston: Routledge and Kegan Paul, 1984.

CARE, Bangladesh, 'Attitudinal Study of the Preventive Maintenance Programme Pilot Schemes', Dhanmondi, Dhaka, 1983.

M. Chen and R. Ghaznavi, 'Women in Food-for Work: The Bangladesh Experience', Dhaka: Bangladesh Rural Advancement Committee (mimeo), 1977.

Stefan de Vylder, *Agriculture in Chains, Bangladesh: A Case Study in Contradictions and Constraints*, London: Zed Press, 1982.

G.W. Edgren, 'Spearheads of Industrialisation or Sweatshops in the Sun?' ARTEP-ILO, PO Box 2-146, Bangkok, 1982.

Felicity Edholm, Olivia Harris and Kate Young, 'Conceptualising Women', *Critique of Anthropology* 3(9/10): 101-30, 1977.

Diane Elson and Ruth Pearson, ' "Nimble Fingers Make Cheap Labourers". An Analysis of Women's Employment in Third World Export Manufacturing', *Feminist Review* 7 (Spring): 87-107, 1981.

Shelley Feldman, 'The New Industrial Policy and Women's Wage Work in Bangladesh', Field notes from Fulbright Grant No. 83-006-IC, March through October, 1984.

—— 'The Transformation of the Domestic Economy: New Household Labour Relations in Bangladesh', Paper presented at Association of Women in Development Conference, 'The Global Empowerment of Women', Washington, DC, November, 1989.

Shelley Feldman and Florence McCarthy, 'Purdah and Changing Patterns of Social Control Among Rural Women in Bangladesh', *Journal of Marriage and the Family* 45(4): 949-59, 1983.

Rural Women and Development in Bangladesh: Selected Issues, Oslo: NORAD, Ministry of Development Cooperations, 1984.

F. Fröbel, J. Heinrichs and O. Krey, *The New International Division of Labour*, Cambridge University Press, 1980.

Annette Fuentes and Barbara Ehrenreich, *Women in the Global Factory*, Boston: South End Press, 1984.

Grameen Bank, 'Annual Reports, 1980-85', Dhaka: Grameen Bank, 1985.

Charles D. Gray, 'Protection or Protectionism', *Far Eastern Economic Review* (13 September): 15, 1990.

N. Heyzer, ed., *Daughters in Industry: Work, Skills and Consciousness of Women Workers in Asia*, Kuala Lumpur: Asian and Pacific Development Centre, 1988.

Hameeda Hossain, Roushan Jahan and Salma Sobhan, 'Industrialisation and Women Workers in Bangladesh: From Home-Based Work to the Factories', in N. Heyzer, ed., *Daughters in Industry*, Kuala Lumpur: Asian and Pacific Development Centre, 1988, pp. 107–135

Mahabub Hossain, 'Credit for the Rural Poor, The Grameen Bank in Bangladesh', Dhaka: Bangladesh Institute of Development Studies, Research Monograph, 1985.

F. Tomasson Jannuzi and James T. Peach, *The Agrarian Structure of Bangladesh: An Impediment to Development*, New Delhi: Sangam Books, 1982.

M.E. Marum, *Women at Work in Bangladesh*. Dhaka: USAID, 1982.

Florence E. McCarthy, 'Patterns of Employment and Income Earning Among Female Household Labour', Dhaka: Ministry of Agriculture, 1980.

T.G. McGee, 'Labour Mobility in Fragmented Labour Markets, the Role of Circulatory Migration in Rural-Urban Relations in Asia', in Helen I. Safa, ed., *Third World Countries*, Delhi: OUP, 1982, pp. 47–66.

June Nash, *et al.*, ed., *Women, Men and the International Division of Labour*, Albany: State University of New York Press, 1983.

New Nation, 'EPZ: Only Four Industries Approved,' *New Nation*, 1(8), 6 November 1982.

Ruth Pearson, 'Female workers in the First and Third Worlds: The Greening of Women's Labour', in R.E. Pahl, ed., *On Work: Historical, Comparative and Theoretical Approaches*, New York: Basil Blackwell, 1988, pp. 449–66.

Brian Phelan, *'Made in Bangladesh'? Women, Garments and the Multi-Fibre Arrangement*, London: Bangladesh International Action Group, 1986.

Charles Tilly, 'Transplanted Networks', New School for Social Research (October), 1986.

UNDP (United Nations Development Programme), 'Bangladesh Agriculture: Performance, Resources, Policies and Institutions', Dhaka: Bangladesh Agriculture Sector Review, 1988.

Ben J. Wallace, *et al.*, *The Invisible Resource: Women and Work in Rural Bangladesh*, Boulder: Westview Press, 1987.

8

Women in Development and Politics: The Changing Situation in Sri Lanka[*]

AMITA SHASTRI

The record of Sri Lanka relating to women is remarkable in several respects. Sri Lankan women gained the right to vote as early as 1931, just a decade after women gained this right in the United States and three years after women in Great Britain. They were the first in Asia to gain this right. All the available evidence indicates that they have since voted in almost equal proportions to their male counterparts which, again, is unusual in view of the gender gap that has been the norm in most Western societies. In 1960, Sri Lanka made history by becoming the first state to have a woman head a democratically elected government. Alongside these achievements, Sri Lankan women also experienced a dramatic improvement in their circumstances in the past half century. There was a progressive narrowing of sex differentials in educa-

[*] Earlier published in *Journal of Developing Societies* (Toronto), 1991–92. My thanks to the editors of the journal for permission to print it here. An earlier version of this paper was presented at the conference 'Bridging Worlds: Women in South Asia', University of California, Berkeley, January 1988.

tion and literacy, a rise in the age of marriage, and sharp declines in fertility rates: these were unusual by Third World standards and approximated more closely to the pattern observed in developed western societies. Yet Sri Lanka also continued to conform to the worldwide pattern of unequal participation by women in formal public, economic and political arenas.

This anomalous situation has evoked a variety of responses. Nationalists have taken justifiable pride in Sri Lanka's achievements but coupled it with the less justifiable presumption that these were due to the inherent liberal temper and genius of the Sri Lankan people (or, alternatively, its elite). This view is reinforced by those who enthusiastically contend that the liberal temperament is nurtured by the central tenets of Buddhism and the local Sinhalese culture. While cultural variables may be partially valid explanations for action, the larger question relating to the selection and sustenance of these liberal elements rather than their opposite (as in other erstwhile Buddhist societies like Kampuchea and Vietnam) remains unaddressed by such explanations.

Liberal social scientists attribute the favourable situation of Sri Lankan women to the ongoing process of 'modernization' in the twentieth century. The process of modernization, with its increase in communications, literacy, urbanization and industrialization, provided diverse and increasing opportunities for economic participation by women. This enhanced the prospects for their political participation and access to formal political power (Jayaweera in Fernando and Kearney, 1979; Kearney, 1981). By and large, these scholars are optimistic about the prospects of women for the future, a view which largely also informs popular thinking about the subject. To quote Kearney, 'It can be presumed that continuing modernization will lead to greater flexibility in the life pursuits available to women and to improved life circumstances that will enhance the autonomy and sense of personal efficacy of women, with the result that a steadily growing number of women will enter the political life of the nation' (Kearney, 1981: 746).

To my mind this view is unduly sanguine and holistic. In this article, I evaluate the anomalies in the situation of Sri Lankan women in a historical perspective and elucidate the somewhat fortuitous manner in which the contemporary situation evolved. I then relate the changing role of women in the public sphere in Sri Lanka in an integral manner to changes in the overall socio-economic structure and the opportunities it offered to different classes of women. In closing, I look at recent changes and find cause for concern about the prospects for women's increased participation in the forseeable future.

THE PRE-INDEPENDENCE PERIOD

Traditionally in Sri Lanka, as elsewhere in the world, the location of women in the home was integrally connected to the imperatives commanded in their lives by the biological facts of pregnancy, childbearing and lactation. These assumed overriding and determining importance for the mass of women in the context of high fertility and mortality rates and the low level of productive forces characteristic of pre-capitalist agrarian societies. The liberating influence of Buddhism and the local marriage customs provided a greater freedom and reciprocity of rights and obligations to Sinhalese women. Yet, at no historic period did they enjoy a truly equal position and status with men (Jayawardena, 1986: 110–15). The subordination of women was enforced more strongly amongst upper caste orthodox Tamil Hindus residing in the north-eastern parts of the island, and through the strict seclusion of women favoured by the Muslims in the east. These cultural distinctions between the ethnic groups continue to remain important till today.

The importance of the right birth, wealth and family connections within the pre-colonial patriarchal social structure excluded all but a miniscule proportion of women from potential access to political power. Such power was exercised by upper class women only in the absence of a suitable male relative. However, as mothers, wives and daughters of great kings, a number of women on the island made their contributions to the religious and political events of their time. From the 'demon' queen Kuveni (who ruled Sri Lanka when Prince Vijaya arrived on the island and married him) to Sugala who led the armies of her kingdom into battle, Sri Lankan historiography is embedded with stories of these and other equally famous queens (Jayawardena, 1974: 17). They remain to this day potent symbols of female power in popular mythology and historiography.

The possibilities for even the occasional exercise of power by women declined with the spread of European colonial rule through the force of arms—a process completed in 1815 when the whole island was taken over by the British. The organization of the colonial state along military-bureaucratic lines and centralized executive decision-making explicitly excluded women from within its portals for over a century. It is remarkable that though the Municipal Council of Colombo (the major city and capital of the island) was formed in 1865 as the first elective institution of government on the island, the first woman member only entered it in 1937 for the brief spell of one year. The elective principle was introduced to the national State Council in 1912 on the basis of a very restricted male franchise. The first woman entered it in 1931 only when universal franchise was introduced.

However, the growth of the plantation export economy in the fertile south-western part of the island after the mid-nineteenth century brought important structural changes in its wake and affected women in several ways. With the expansion of education in the nineteenth and twentieth centuries, a number of urban middle class women entered teaching, nursing and secretarial work (Jayawardena, 1986: 118–122). Concomitantly, the situation of the peasantry progressively deteriorated so that rural lower-class women and girls, who had earlier laboured as unpaid family, artisanal or subsistence workers, were forced to work as unskilled wage labour in the paddy fields and the newly established plantations and factories. The socio-economic position of this group of women and those continuing on unviable subsistence holdings continued to decline into the 1930s (Grossholtz, 1984: 96, 119).

Though excluded from the portals of formal power, Sri Lankan women began informally to get increasingly involved in the political events of their time. The struggle for democratic rights which began with the Buddhist revival of the 1880s in Sri Lanka, the temperance movement and early nationalist agitations encouraged a successively more activist public role by upper and middle class women. The earliest singular participation by these upper class women was as 'hostesses'. Like similar women in the US, UK, and India, they used the social influence, financial resources, the large estates and buildings of their families to house politically active (often foreign) guests; and host meetings in support of causes favoured by them. Other educated women, several of whom were Tamil, became prominent in the cultural field in the early decades of this century as novelists, writers and poetesses (Jayawardena, 1974: 19–20). Their increasing participation in public life was stimulated by events abroad, especially the broad-based nationalist struggle taking place in neighbouring India and the foment of radical activity in Britain (ibid).

Following close on the heels of these earlier women, some of the newly-emergent professional and middle-class women became more conscious and politically active. A few of them joined nationalist organizations like the more radical Young Lanka League (formed in 1915), and the moderate Ceylon National Congress (formed in 1919). They also formed their own organization, the Mallika Kulangana Samitiya, allied to the Congress. Using the Samitiya, they lobbied with limited success in the Congress in 1925–26 for a recommendation for a restricted female franchise (Metthananda, 1981: 67–68). In 1927, the Women's Franchise Union was formed to give evidence before the liberal three-person Donoughmore Commission which was sent to reform the constitution. The Union argued that women had a special contribution to make on matters such as health, education, and social welfare as these pertained particularly to women and children.

It is noteworthy that members of the Women's Franchise Union were willing to accept high literacy and property qualifications if female suffrage were conceded. As upper-class or educated middle-class women, many of them wives and mothers of eminent nationalist politicians, they were by and large in agreement with their male counterparts and argued for a qualified franchise before the Donoughmore Commission.[1] However, when the Donoughmore Commissioners recommended a higher age restriction for women (thirty years rather than the twenty one years recommended for men) with no property or literacy qualifications, the Union argued strongly for the award of franchise to the younger group of women on the same terms as men (Metthananda: 68–69).

In this context, it is significant that no movement comparable to the suffragette one in Britain and US with strong links to middle and lower-class working women developed in Sri Lanka. Indeed, it could not do so since the process of organization of the working class itself was in its incipient stages when universal franchise was awarded to Sri Lanka in 1931. The more radical political parties and trade unions such as the Ceylon Labour Party, the Lanka Sama Samaja Party, the Communist Party and the Ceylon Indian Congress were formed only in the mid-twenties and thirties. Instead, universal adult suffrage (for those above twenty one years) was introduced at the behest of the colonial authorities who considered it necessary to balance and broad-base the greater powers for self-government that were, implicitly, going to be conceded to the Ceylonese elite by the Donoughmore Constitution of 1931. Female suffrage was thus gained in Sri Lanka at this early date with hardly any agitation and over a fairly short period when qualifications for women were more restrictive than those for men.

The introduction of universal franchise made it necessary for the national elite to compete for popular votes with the leaders of the increasingly active working class movement. While the wealthy landowning and business elite could mobilize traditional rural loyalties and divisions of ethnicity, locality, and caste to work in their favour, the need to sustain that loyalty and to win a following in the more politicized, sceptical and volatile urban population encouraged the elitist legislators to compete with and attempt to outmanoeuvre the leftists through the award of state patronage and welfare benefits.

As a result, the other major action taken in the pre-independence period that was to have a profound impact on the situation of women—the state provision of a generous array of social welfare programmes—was also undertaken with little demand from women as a group. Following reports on the worsening condition of the peasantry in the rural areas and to maintain support for the war effort, a paternalistic upper-class leadership legislated programmes providing free health and education facilities, and free food rations to the people. This was financially possible in the background of a booming plantation economy of the pre- and post-Second World War years and had an immense impact on the physical well-being of the population at large, and women in particular, as we will see in the next section.

Thus, with the onset of capitalist development and growth in the economy, the situation of women in Sri Lankan society began a process of historic change. Women became increasingly involved politically in the larger nationalist and labour movements of their time and it is as part of those movements that they made their most notable gains and contributions. Efforts to focus on and organize for issues pertaining solely to women and their social position remained limited and subordinate in character.[2]

THE WIDENING BUT DIFFERENTIAL IMPACT OF DEVELOPMENT

It would be useful in this section to examine the specifics of the change in the situation of Sri Lankan women in the past century, especially in the latter half since the institution of the welfare measures. These measures were continued in the post-independence period despite a chronic deterioration in the terms of trade and a slow rate of economic growth after 1956. Protective labour legislation applicable to the organized sectors of the economy was also introduced in the 1950s. However, as we will see, the remarkable gains made by women remained limited in important respects.

As a consequence of the various health and welfare measures, there was a dramatic improvement in the prospects for physical survival of Sri Lankan women. It was particularly remarkable for its improvement beyond that of men in a resource-scarce Third World society where men are traditionally provided preferential access to family resources. Expectation of life at birth for females more than doubled in the course of half a century from an average of thirty-one years in 1921 to seventy-three years by 1987 (World Bank, *World Development Report 1989*: 184). The expectation of life surpassed that of males by five years in the same period whereas it had earlier fallen short by two years!

Equally remarkably, female literacy climbed steadily for a hundred years from some 3 per cent of the female population in 1881 to 44 per cent by 1946 and to 82.8 per cent by 1981. In that year, 85.3 per cent of urban females above ten years of age and 78.1 per cent rural females were reported literate. (Department of Census and Statistics, *Labour Force and Socio-Economic Survey* [or SES] *1980–81*: 120). The average rate of female literacy of 82.4 per cent in Sri Lanka in 1981 is in striking contrast to the 33 per cent rate that existed in South Asia as a whole. Moreover, the introduction of universal free education from kindergarten to university in 1944 encouraged an increasing number of women to complete school and even university in proportions comparable to their male counterparts. From constituting only 10.1 per cent of those seeking higher education in 1942, the proportion of women reached 42 per cent in 1972, a figure fairly close to their overall proportion in the population. In regional terms, female literacy was highest in Colombo and the more developed south-west of the island and lowest in the less developed rural parts of the Eastern Province (Kearney and Miller, 1987: 53). The more general PQLI index for Sri Lanka was estimated to be 82 in the years

1970–1975,—a figure surpassing those of the other Third World states, and approaching the indices of 90–97 characteristic of the most developed countries.

The gains made in health and literacy worked with other factors, often in a contradictory ambiguous manner, to redefine the private and the public spheres in different segments of the society. They had an impact on the norms and values relating to marriage, children and the family, and the prospects of employment outside the home. Initially, longer life expectancy and lower death rates contributed to a high rate of growth of population and the tendency towards larger families. The tendency to increased family size was closely followed by a drop in the participation rate of women in the labour force from 32.2 per cent in 1946–1953 to 23.7 per cent between 1953–1963 as more women remained at home to raise their families. However, no doubt due to the rising rates of literacy and economic pressures at home to increase incomes, the participation rate rose back to 31 per cent in the sixties, a figure close to the world mean (UN, *Socio-Economic Development and Fertility Decline in Sri Lanka*, 1986: 63). This was accompanied by a rise in the age of marriage for women from 20.7 years in 1946 to 24.4 by 1981. And, as the decline in birth rates show, there emerged with a time lag a limited, and even declining, family size. The decline in fertility was initially only partially due to increased reliance on contraceptive devices per se (C. H. S. Jayawardene, in Fernando and Kearney, 1979: 49–51; Gunasinghe, 1977) but this apparently changed after the mid-seventies with a significant increase in the practice of family planning. A survey done in 1975 found that only 18.8 per cent of respondents used modern contraceptive methods (UN, 1986: 89), but by 1985 the number of women of childbearing age who reportedly used contraceptive devices had increased to 62 per cent (*World Development Report 1989*: 184).

Though women had equal access to education, sex-based diversifications in curriculum in schools and traditional cultural attitudes influenced girls against joining 'masculine' areas such as technical studies. An overwhelming proportion of them opted for the arts, law, commerce, and the social science options—subjects which, unfortunately, offered poorer employment prospects in the longer run. Similarly, though highly educated, due to persisting sex-role stereotypes, 46 per cent of women continued to be engaged exclusively in housework as against less than 2 per cent of the men. Another one-fifth of the female population were students (SES 1980–81: 15).

Consequently, only a small, though fast increasing, proportion of women were gainfully employed outside the home. As late as 1980, women formed only 27.4 per cent of the economically active population (ibid). Reflecting the national distribution, over half these women were engaged in agriculture and related activities, about one-third were in the service sector, and about 15 per cent were in manufacturing. There were four industries which had a particularly high proportion of women workers: plantation agriculture (47.3 per cent), the manufacture of textiles and wearing apparel (63.5 per cent), education and health services (43.1 per cent), and domestic service (59.8 per cent) (*Census of Population, 1971*, vol. 2, part 2).

Characteristically, as in most other countries, the large majority of women found employment in labour-intensive, low-income jobs. A disproportionate number of them were also engaged in work in the informal, unorganized sectors of the economy where labour laws did not operate. Available data indicates that while more women have moved into professional and technical fields recently, there is no change in the meagre proportion of 0.1 per cent of them that occupy administrative or managerial positions.[3] A greater proportion of women continued to be employees and unpaid family workers in contrast to men who were more likely to be employers or workers on their own account (*SES 1980–81*: 21–2). Consequently, as can be expected, the large majority of women remained poorly placed when it came to developing their individual competence, initiative, organizational skills, and networks in the public sphere for both economic and political purposes.

The slow rate of economic growth and the high rate of unemployment characteristic of Sri Lanka after the mid-fifties had a particularly pernicious impact on women. The unemployment rate rose steeply for women to become double that of men: from 9.5 per cent for men and 12.4 per cent for women in 1959–60, it increased to 11.4 per cent for men and 21.2 per cent for women in 1969–70 and continued to grow to over a quarter of the labour force into the late seventies. Despite this, unemployed men continued to be the main beneficiaries of unemployment relief schemes (Jayawardena and Jayaweera, 1986: 10).

The material situation of Sri Lankan women was reproduced and reinforced by traditional patriarchal ideology. Women were viewed as secondary earners due to their poorly-paid subordinate status and society continued to place the responsibility and burden of maintaining a smooth and stable family life primarily on women. Women were (and are) required to be accommodating and docile, less capable and aggressive than men in the private sphere. This had its obvious spill-over effects on behaviour patterns approved of in the public sphere—in the work place and in public affairs. This was despite the fact that a growing proportion of female workers consisted of married women. Reportedly, by the mid-seventies, married women constituted 51 per cent of all female workers (*Report*, 1979: 447). The availability of the cheap labour of lower-class men and women as servants alleviated the situation for the better off amongst the middle-class and professional women. Still, it continued to be the case that the large proportion of women remained seriously restricted in any role they could potentially play in political and public life.

LIMITED POLITICAL PARTICIPATION

The dominant forms of political activity by women in the post independence period in Sri Lanka was conducted through the accepted vehicles of political parties, elections and the parliament. It would be true to say that till recently issues of relevance to women *qua* women remained submerged amongst the larger issues related to economic growth, structural reform, and ethnic competition that were dominant at the political level. Electoral mobilization contributed to increasing group cleavages along lines of class, region and ethnicity. The trade union movement, likewise, got fragmented along ethnic, sectoral and party lines. Estate Tamil workers, amongst whom women were particularly numerous, were disenfranchised within two years of independence and large numbers of estate Tamils were repatriated to India or remained without the vote into the 1980s. In such a milieu, women remained subordinate to these divisions and engaged in political activity within them.

The political importance of women in numerical terms in a majoritarian democracy caused all major political parties to organize women's wings or front organizations to seek out and woo them as voters (See *Report*, 1979: 569). At the same time, the emergence of an issue-oriented two-party democratic parliamentary system in Sri Lanka after 1956 served to draw an increasing proportion of women into the electoral process. The deteriorating economic context also served as a catalyst to women's electoral participation. As housewives and mothers, women were acutely affected by the problems of budgeting for their households in the context of soaring living costs, a scarcity of essential commodities, cuts in the rice subsidy by successive troubled regimes, and high rates of youth unemployment. Unfortunately, it is difficult to say anything at the macro or aggregate level about specific rates of female voter participation and registration because such data broken down by sex are non-existent for Sri Lanka. The fact that the overall rates of voter participation rose rapidly from 55 per cent in 1947 to reach almost 87 per cent in 1977 indicates that the rates of participation by women must necessarily have been very high by the end of this period—about 75–80 per cent. By all accounts, the differentials in voting by sex were very low. A study in which women were sampled from four different areas of the island (urban and rural), reported a voting rate for women ranging from 77 per cent to 99 per cent. A large proportion of the women surveyed, moreover, claimed to have made individual choices of who to vote for independent of their husbands.[4]

Very few women aspired to occupy the more competitive and demanding leadership positions in the political sphere. According to one estimate, women constituted between one-quarter and one-third of the membership of major parties (Jayaweera, in Fernando and Kearney, 1979: 166) but constituted a much smaller proportion of party officers or leaders. A computation made in 1979 found that 10 per cent of the members in the United National Party's highest decision-making bodies, 7 per cent of the Sri Lanka Freedom Party's and 2 per cent of the two left parties were women (*Report*, 1979: 569). Likewise, at the end of January 1979, the Central Federation of Trade Unions consisting of 17 affiliated unions had only one woman on its 32-member executive committee and 2 women members on its 132–person general council (ibid: 578). It has been observed that even in organizations where women comprised the majority of the membership as in the nurses, teachers and typists' unions, the elected leaders were generally men (ibid: 166).

While peaceful organizational activity constituted the more accepted medium for women, the high rates of unemployment amongst the youth engendered alienation which was increasingly manifested through radical political activity. Perforce, women also emerged as colleagues and helpmates in clandestine and dangerous ventures. They formed 2 per cent of those arrested as JVP insurgents in 1971.[5] More recently in the 1980s, Tamil women who had previously acted as information and support agents to male Tamil activists, began to physically train for and participate in the operations as militants.[6]

At the elite level, since women had not established themselves in roles and institutions of power and patronage on a routine footing with upper and middle class men, women who could logically aspire to elected political office were few in number. Given the traditional class and status-conscious society, the family into which women were born and the opportunities that emanated therefrom became critical factors in determining whether an individual woman chose a political career. I agree with the assessment that in Sri Lanka, as in India, the considerable prominence a very few women have attained in politics has undoubtedly conveyed a distorted impression of the active role of women in general in these societies (Kearney, 1981: 733; Katzenstein, 1978: 473–86). Only between 1 per cent and 4 per cent of the members of Sri Lanka's parliament have been women.[7] This figure is not substantially different from the proportions of women found till recently in most national legislatures constituted through competitive elections (Currell, 1974: 26). The proportion of women as candidates has also been extremely low and fluctuating (between 1 per cent and 3 per cent), giving us little reason to conclude that the proportion of women in political offices is showing any definite long term tendency to improve—as Kearney suggests (Kearney, 1981: 745, Table 1). In addition, of the 6 MPs appointed to every parliament between 1947–1977 to represent important but electorally unrepresented communities or constituencies, not one was a woman. Similarly, a mere 6 women served in the weak upper chamber, or senate. Of these, 2 were elected and 4 were appointed. Unlike their male counterparts who were drawn on the basis of public service and eminence in various walks of life, the women senators were all from an elitist urban background and had engaged in social work.

Due to the very small number of women who entered parliament, and the even smaller number who served consecutive terms, a miniscule number of women served as cabinet ministers. Besides Mrs Bandaranaike, only 3 other women had been in the cabinet till 1981. Of these, 2 previously served as parliamentary secretaries (Kearney, 1981: 741). Of the 22 district ministers appointed for the first time in 1978, only one was a woman.

Not surprisingly, then, the women who entered did so at their family's behest, which implicitly means with the active encouragement and support of important male relatives. Most often, they came into politics explicitly as male equivalences and were expected to support the policies of their menfolk. This is very similar to the situation described for UK and USA, especially for the pre-war period. Of the 17 women who were elected to parliament after independence in 1947, more than half were elected, or first elected, to parliament as replacement for a male relative who had died or whose election had been declared void (Kearney, 1981: 737).

Indeed, it is pertinent to highlight that the best-known of them all, Sirimavo Bandaranaike, succeeded her assassinated husband to become the first woman prime minister in the world in 1960. This was done, interestingly enough, without even contesting the 1960 election! A procedure unusual to the British parliamentary system was adopted on her party's victory at the polls (a victory in large part due to her effective campaigning). She was named prime minister by the governor-general; and then recommended her own appointment to the senate—the upper chamber—from which she led her majority party in the lower chamber for almost five years! One of the qualifications which was considered important and favourable at the time of her nomination was the fact that she was a descendant of the exclusive traditional high-caste landowning Kandyan aristocracy and through her marriage was linked to one of the most substantial low-country landowning families. Thus, one of the apparently most dramatic achievements of women in Sri Lanka, that of having the first woman prime minister of an elected democratic government, was done through the most traditional and conservative routes that have historically been available to women—of nomination as a male equivalence by a set of male party influentials!

This is not to derogate the accomplishments of these women as individuals. It needs to be added that, like male politicians of their class, having once entered the political arena, to be successful they had to swiftly 'professionalize' themselves to build upon their initial entry into politics. Mrs Bandaranaike did this in ample measure and, it might be added, to a lot of dismay. Along with reliance on male kin, she developed and demonstrated an uncommon sense of political efficacy and realpolitik. She served as prime minister for nearly twelve years all together—longer than any other person—and as leader of opposition for five years. As the leader of the United Front government in the 1970s, she concentrated and exercised an amazing amount of power, often arbitrarily. She has remained at the helm of the SLFP for thirty years now, despite the depradations of the ruling UNP on her civic rights and political position after 1977, and continues to remain a force to contend with.

In contrast to Britain, where the suffragette movement encouraged women to emerge as independent candidates, most often espousing women's interests specifically, independent women candidates were rare in Sri Lanka. Instead, three-fourths of the women candidates had stood for the more durable and well-organized parties. Only 8 out of 78 women candidates in the post-independence period till 1978 were independents and another 8 stood as candidates of small short-lived parties (Kearney, 1981: 734 –5).

Given the predominant tendency towards elitism and status consciousness, it is not surprising to find that women were even more poorly represented in local government bodies than in the national legislature. Predictably, this under-representation was more marked in town and village councils than the more prestigious municipal and urban councils (for details see Kearney, 1981: 742).

Thus we find that while a few women could play a highly visible role at the national level in Sri Lanka, the large majority remained in a subordinate status at home, at work and in politics.

THE CHANGING SITUATION

The failure of previous governments to effectively promote an adequate rate of economic and industrial growth and to curb the rising levels of unemployment led to a reversal of the earlier nationalist policies in 1977. The new economic policy espoused by the UNP regime encouraged domestic capital and opened the economy to foreign capital and trade, along the model of Singapore and the other East Asian Newly Industrialized countries (NICs). From the data available it appears to have had a significant impact on the situation of women in Sri Lankan society and, in many respects, it provides cause for concern. As we will see below, the situation has been made worse by the prevalent violence and conflict. The changes have, consequently, led to new modes of organization by women.

One of the noteworthy features of the new pattern of development has been its high absorption of female labour power as cheap labour. The educated, unemployed status of Sri Lankan women has been widely advertised in the international business media as a lucrative basis for profitable investment in low-skill labour-intensive lines of industrial production. Women have been employed in increasing numbers in assembly-line production in the garments, leather-goods, electronics, jewellery-making and similar export industries encouraged in the newly established free trade zone (or FTZ) near Colombo. According to provisional estimate, of the 41,614 workers employed in the FTZ by 1986, around 34,000 were employed in apparel industries, producing textiles and leather products (Greater Colombo Economic Commission). The overwhelming proportion of these employees—some 80 per cent —were women. Women have also become employed as receptionists, secretaries, tourist guides, housekeepers, waitresses, bus-conductors, and salesgirls in the urban tourist, transport, trade, and banking sectors encouraged by the new regime. A number of women have emerged as entrepreneurs in the informal sector to own and manage lodgings and restaurants, and make or sell handicrafts to tourists. Yet a large number of others, presumably of rural landless origins, have become construction workers in the massive Mahaveli Development Project, a major irrigation and power development scheme in the centre of the island. Similarly, others have become construction workers in various schemes for urban development and low-cost housing across the island.

The new 'open' policy has also indirectly encouraged other trends in women's employment. Lacking adequate opportunities in the country and compelled by economic need, many women from the lower class, often from otherwise conservative Muslim backgrounds in the Eastern Province, have emigrated to West Asian countries to become domestic maid-servants. Though the work is low-paid and menial, it demands few skills and offers the highest differential to the pay available in Sri Lanka across all categories of work—a differential of more than thirty times (Ruhunage in Bandarage, 1987: 70). In the peak year of 1983, some 30,000 women went to West Asia, outnumbering male migrants. An unspecified number of others joined the expanding tourist and entertainment industry at home as prostitutes and show-girls.

As a result of the new opportunities, the percentage of the female labour force employed increased from 75 per cent to 79 per cent by 1982. There was an especially sharp drop in the unemployment in the FTZs and the urban informal sector, so that urban female unemployment rates approached the lower rates of rural unemployment.

The above changes were also reflected in a restructuring of the female labour force in various industries. While a declining proportion of women, like men, remained in agriculture, an increasing proportion joined the urban manufacturing sector. The data available indicate that in the 1970s as a whole, the proportion of the female labour force employed in manufacturing increased from 14 per cent to 20 per cent (*Census of Population*, 1971 and 1981), a trend which no doubt continued into the 1980s. Between 1979 and 1984, there was a slight favourable shift in the small proportion of women at different levels of the administrative services, academic and education services (Jayawardena and Jayaweera, 1986: 74–8).

The emphasis in official policy in favour of increasing foreign and private investment and profits to encourage the growth of the economy has been accompanied by a series of laws controlling union and open political activity. These have worked to the detriment of the women workers employed in the new manufacturing and service industries. The UNP government has not only passed stringent legislation limiting the rights of workers to strike but the rights of workers have been rendered inoperative in the FTZ and other industrial units operating under its regulations. The age for employment has been lowered, so that any person above 14 years of age can be employed in a FTZ office or factory. This has encouraged the hiring of young, docile, unmarried female labour, who are preferred over the generally more assertive male labour. Both are preferred to married female labour whose work tenure is open to interruptions of pregnancy and household responsibilities and to whom maternity benefits would fall due in organized sector employment. The women or girl workers are paid a little more in the FTZ than elsewhere in the country but they also have to work harder under strictly hierarchical and regimented conditions. They have no job security and no compensation in the event of accidents. Prohibitions on night-work were removed in 1982 and workers are often compelled to do overtime (*Kantha Handa*, 1983).

The liberal spending of foreign funds and public borrowings by the government after 1977 has resulted in a high rate of inflation which has eroded the real incomes of significant sections of workers and fixed-income earners. This has been worsened by government cuts in consumer subsidies in food, fuel, health and education. The cost of living as measured by the Colombo Consumer Price Index, which is the only official indicator available and which tends to understate price increases, rose on an average by 12 per cent per annum between 1981 and 1986. According to official sources, minimum wage rates declined till 1981–2 (1978=100) but recovered in the following period and even showed a gain over the 1978 figure by 1985–6. The decline in wages was, however, particularly sharp in industry and commerce and did not recover even by 1986. The corresponding index declined to 79 by 1984

(1978=100) and had risen only to 89 by 1986 (Central Bank of Ceylon, or CBC, *Annual Reports*).

Even though the proportion of the female labour force that was employed increased, the proportion of women to be found in paid employment has continued to remain very low—equal to one-third that of males (CBC, *Consumer Finances and Socio-Economic Survey 1981–82, part 1*: 204). Instead, the limited available data indicate that differentials in income between male and female income receivers have widened. The ratio between median incomes of males to females rose from 2.24 in 1978–79 to 2.36 by 1981–82 (ibid), highlighting the fact that women were increasingly being segregated into lower paying jobs and failing to take advantage of the opening opportunities on an equal footing with their male counterparts.

It is also worth noting that though there was a greater drop in unemployment amongst females between 1978 and 1981–82, the unemployment rate for females remained two and a half times as high as that of males. In 1981, while 13.2 per cent of men were unemployed, the corresponding figure for women was 31.8 per cent (*Census 1981*). Of as much concern is the fact that 34.5 per cent of the unemployed females had passed the G.C.E.(O.L.) examination as against only 19.6 per cent unemployed males; 9.5 per cent of unemployed females had passed the advanced level school examination and above as against 3.7 per cent of males.

At the structural level, it seems that the high rates of unemployment have worked with insidious effect to encourage a significant proportion of women to withdraw from their search for new opportunities for economic advancement. According to official data, a rising proportion of the female population of working age has opted to remain economically inactive. In contrast to 1970 when 42.6 per cent of females were occupied with housework, in 1980, 45.9 per cent females were similarly occupied.

Of additional concern is the fact that the high rates of unemployment of educated youth seem to have discouraged the upcoming generation of young women from seeking education as a means to economic independence and mobility. This underlines an earlier trend observable amongst those seeking higher education—the decline in the proportion of women from the all-time high of 42 per cent in 1972 to 39 per cent by 1976. In 1987, there were reportedly 109 females to 100 males in secondary school but only 93 females to 100 males at the primary school level (*World Development Report 1989*: 184).

The general effects of the new economic policy can therefore be judged to have been deleterious to the overall position of women. Women form a large proportion of the new migrant labour streams and experience the attendant strains. A common characteristic of the new avenues of employment for women has typically been an increased subordination in the production or service hierarchy. The large majority of women have thus given up the personalized domination by kinfolk and men at home for the impersonal tyranny of the shop floor—in the FTZ, as construction workers, as maids, as prostitutes. In their new avenues of employment they are rewarded with low pay for long hours of work, often accompanied by abuse, exploitation, insecurity, stress, and poor health. At the same time, the government is increasingly shifting the burden for the reproduction of labour back to the private sphere. Faced with declining real incomes, with less support extended by government policy, and declining prospects for changing the same through organized collective action by labour, the burden of the new development strategy is being borne inordinately by the female section of the population. Attempting to balance their responsibilities at home with the demands made by their jobs, women are too often faced with deteriorating relationships at home with their families and children. They are unable to fulfill the traditional ideals of wife and motherhood and yet are judged against those standards. As a general norm so far, the prevailing cultural constructions of gender and male privilege have evoked little support from the men in their families for a more equitable division of labour and responsibility in the private domestic sphere.

To add to the gravity of the situation, there has been a slowing down in the rate of growth of the economy in the late 1980s. Construction activities on the island began winding down after 1986 as the major projects were nearing completion and there was a shortfall in capital investment funds. There has been a sharp drop in opportunities for employment in West Asia in recent years, and especially in recent months due to the Gulf crisis. These trends have been made worse by the severe ethnic conflict that has been raging in the island since 1983 in the north and east and the JVP insurgency in the south; both of which struck the tourist industry particularly hard. The result has been marginal or negative real growth rates of the economy in the latter part of the 1980s (CBC, *Annual Reports*). Estimates of unemployment for the years 1989–90 ranged from 18–20 per cent (Economist Intelligence Unit, or EIU, *Country Profile: Sri Lanka 1989–90*: 10). The unstable domestic situation has also contributed to lowered expenditure on public investment and to a diversion of resources to the armed forces. Whereas the total expenditure of the central government as a proportion of the GNP rose from 25.4 per cent to 32.4 per cent between 1972 and 1987, the proportion allocated to defense also increased from 3.1 per cent to 9.6 per cent for the same years (*World Development Report 1989*: 184).

The rising spiral of violence and coercion between the ruling party backed by the army, police, and its para-military squads and popular youth groupings like the Tamil militants and the Sinhalese JVP[8] have left little room for normal electoral and parliamentary processes within which women in particular find room to organize and expand influence. Moreover, given the acuteness of conflict between parties in Sri Lanka today, it seems unlikely that the hoped for decline in political violence and character assassination by competing candidates would take place and more women would come forward as candidates for election, as some analysts had earlier argued (de Silva, in *Report*, 1979: 554). It is worth noting that the use of the new party-list system in local government elections in 1979 and the parliamentary and provincial council elections in 1989 did not increase the proportion of women elected (Kearney, 1981: 742; my observations).

Instead, the contemporary violence-ridden situation has compelled women to organize along new lines on an independent basis. A number of women have stepped forward to work with non-partisan civil and human rights groups such as the Civil Rights Movement, Centre for Society and Religion, Movement for Inter-racial Justice and Equality, and Sarvodaya. Many other urban professional women have supported the growth of a feminist movement in Sri Lanka since the late seventies. The movement has derived considerable inspiration from abroad: from the UN's Women's Development Decade, from the women's movements in the US and other western countries, and the women's movement in neighbouring India. Most notable amongst such organizations has been *Kantha Handa* (Voice of Women) in Colombo which has undertaken research, meetings and agitations to raise public awareness of issues related to women. Lobbying by women led to the creation of a Women's Bureau within the Ministry of Plan Implementation, overseen by the President, in 1978. This was further upgraded in status to an independent Ministry of Women's Affairs in 1983.

In addition, as mothers, sisters and daughters who are not active participants in the conflict but are tragically affected by it, a large number of women have created other issue-specific organizations and groups. These associations have sought to moderate the brutal effects of the conflict by organizing searches for missing persons and soliciting donations and supplies to assist refugees. They have raised issues pertaining especially to women, such as rape by members of the security forces. They have also voiced demands for a rational, just and peaceful solution to the conflict. Amongst the most prominent of these groups have been the Mothers' Front based in Jaffna, the Mothers of Missing Youth in Batticoloa, and Women for Peace in Colombo. Affected by the upheaval of war, large numbers of women have been compelled to stretch their resourcefulness for the survival of their remaining family members in a struggle that has made a mockery of the protective bonafides of the male-dominated society around them.[9] Their direct experience seems to make it unlikely that anything approaching the *status quo ante* can be re-established once the hostilities are over. As evident in popular writings by ordinary women (see Tennekoon, 1986–87), the experience has produced the beginnings of a radical critique of the stratified and oppressive patriarchal society.

CONCLUSION

Thus in Sri Lanka the position of women in the processes of development and politics has varied over time and has struck different classes of women with a differential impact. As we saw in the article, Sri Lankan women made major gains in political and economic rights in the pre-independence period as part of the larger society. In the following decades, participation in the welfare programmes and in the electoral process elevated their quality of life to surprising levels by Third World standards. However, despite this, a disproportionate number of women were only able to gain employment in positions demanding low skills, offering low incomes and opening limited opportunities for economic advancement. A significant proportion of others remained unemployed. Attempting to simultaneously fulfill their prescribed roles at home in a resource-scare Third World milieu, the large majority of women remained in a subordinate status at home and in the workplace. As such, they remained poorly situated to undertake initiatives and commitments in the public political sphere. Similar to the situation in neighbouring India and in other democratic polities, it was a very few upper class women who could gain positions of power and visibility in the political arena.

The new economic policy of the UNP regime has had an important impact on the situation of women for over a decade. The policy has encouraged their incorporation in low-skill labour-intensive lines of production and an erosion of the economic and political rights of labour. There has also been a concerted effort by the government to maintain low wages and cut welfare benefits and subsidies. This has worked with a major conservative impact to shift a larger proportion of the burden for the reproduction of labour to the private sphere, and thus primarily onto women. It has encouraged a noticeable proportion of women to withdraw from the workforce and to neglect higher levels of education. The differentials between the incomes of men and women have also increased. The declining legitimacy of the political system and the high level of conflict and violence in the society in the seventies and eighties has widely affected women, especially in the north and east. It has disrupted their lives and families and threatened their very existence. This has led to a new development in Sri Lankan politics with women increasingly organizing as women to raise demands on issues affecting them as well as the society around them.

Sri Lanka's attempt to emulate the export-oriented economies of the NICs seems difficult to believe in right now. It is possible that if peace was re-established and there were to be an extended period of export-led economic expansion and an increasing demand for labour, the employment, wages and economic situation of the mass of women could improve over the longer term as it did in the four NICs. However, surveying the uncertain world economic situation as well as the conflict-ridden Sri Lankan domestic scene, such a scenario seems unlikely. Rather, it appears that in the foreseeable future a new era of political, social and economic change is to be envisaged in which the experience of women is not likely to be as favourable as before.

NOTES

1. A. E. Goonesinha, leader of the Ceylon Labor Union, was the notable exception at the time when he argued for universal franchise.
2. The All Ceylon Women's Conference, formed later in 1944, and the *Eksatha Kantha Peramuna* (or United Women's Front), 1948, helped

to focus on women's issues to a limited extent: see Jayawardena, 1986: 130–1, 134–5 for details.
3. The foreign service was opened to women only in 1958. The administrative service which was started in 1963, had a ceiling of 10 per cent for women recruits. This was raised to 25 per cent in 1975. Quotas limiting women in these and the clerical and account services were removed by the 1978 constitution.
4. While 72 per cent of women interviewed in an urban area of Colombo claimed to vote independent of their husbands' views, a surprising 67 per cent of those in a Muslim village in the south-west of the island also claimed to do likewise: *Report, 1979*: 594–6.
5. Computed from Obeysekere, 1974: 368. This may possibly be an understatement due to the likelihood of more indiscriminate arrests of men.
6. The case of Nirmala Nithyananthan's breakaway from jail in 1984 is probably one of the more colourful episodes.
7. Reference to the parliament here is to the elected lower house before 1972 and the single-chamber legislature after 1972.
8. To cite just one set of figures for this country of 17 million persons, according to one report, about 30,000 people were estimated to have been killed in 1989 alone. And, as of mid-February 1990, 6,750 suspected subversives were in government custody, and another 2,500 were in rehabilitation camps (EIU, *Sri Lanka: Country Report No. 2, 1990*: 8).
9. In 1981, 17.4 per cent of the households in Sri Lanka were headed by women. The figure was higher than 20 per cent in the districts of Galle and Matara in the south and Jaffna in the north: Women's Bureau, in Bandarage, 1988: 78. The situation is bound to have worsened due to the war.

BIBLIOGRAPHY

Asoka Bandarage, 'Women and Capitalist Development in Sri Lanka, 1977–87', *Bulletin of Concerned Asian Scholars*, 20(2): 57–81, 1988.

E. Melville Currell, *Political Woman*, London: Croom Helm, 1974.

Tissa Fernando and R. N. Kearney, ed., *Modern Sri Lanka: A Society in Transition*, Syracuse, NY: Maxwell School of Citizenship and Public Affairs, 1979.

Jean Grossholtz, *Forging Colonial Patriarchy: The Economic and Social Transformation of Feudal Sri Lanka and Its Impact on Women*, Durham: Duke University, 1984.

Kumari Jayawardena, 'The Participation of Women in the Social Reform, Political and Labour Movements of Sri Lanka', *Logos*, 13(2): 17–25, 1974.

——, *Feminism and Nationalism in the Third World*, London: Zed Books, New Delhi: Kali for Women, 1986.

Kumari Jayawardena and Swarna Jayaweera, *Profile on Sri Lanka: The Integration of Women in Development Planning*, Colombo: Women's Education Centre, 1986.

Mary Fainsod Katzenstein, 'Towards Equality? Cause and Consequence of the Political Prominence of Women in India', *Asian Survey*, 18(5): 473–86, 1978.

Robert N. Kearney, 'Women in Politics in Sri Lanka', *Asian Survey*, 21(7): 729–46, 1981.

Tilaka Metthananda, 'Votes for Women 1923–1931', in K. M. de Silva, ed., *Universal Franchise 1931–1981: The Sri Lankan Experience*, Colombo: Ministry of State, Department of Information, 1981.

Gananath Obeysekere, 'Some Comments on the Social Backgrounds of the April 1971 Insurgency in Sri Lanka (Ceylon)', *Journal of Asian Studies*, 33(3): 367–84, 1974.

'Report on the Status of Women in Sri Lanka', Colombo: University of Colombo, mimeo, 1979.

Serena Tennekoon, 'Sri Lankan Women Demand Peace with Justice', *Connexions*, (22): 8–10, 1986–1987.

Women Workers in the Free Trade Zone of Sri Lanka: A Survey, Colombo: Voice of Women, 1983.

9
Organizations and Informal Sector Women Workers in Bombay

JANA EVERETT AND MIRA SAVARA

INTRODUCTION

This paper examines the role of occupational organizations and women's participation in them among five groups of working class women in Bombay: private building sweepers, domestic workers, subcontract workers, fish sellers and *khanawalis* (women who cook for male factory workers who have come to Bombay alone). These five occupations fall within the informal sector (IS), which can be defined following Portes and Walton (1981) as encompassing all income producing activity outside of formal wage contracts. The sweepers and domestic workers participate in informal sector unions, and the subcontract workers and khanawalis participate in women's organizations. The fish sellers nominally are members of a fish co-operative but do not themselves participate in it. The paper is based on data collected on 200 women workers and five IS organizations in Bombay during 1986–8.

One of the most often quoted characteristics of the informal sector (IS) is the isolated nature of the work and the lack of organization within it. To talk of organizations in the IS therefore seems to be a contradiction, since the very development of an organization seems to take the occupation out of the scope of IS activity into the formal sector. Our point of view is that while IS occupations exist within a context the general nature of which is the lack of organization, sporadic attempts at developing organizations do occur.

Since the development of organizations among IS workers is seen as crucial for improving conditions of work among them, a study of these attempts at organizing could be helpful in developing our understanding of what are the problems and potentials of such organizations, and what are some of the conditions required for a sustainable organization to develop. In addition, what sort of organizations seem to take on women's issues, and to what extent do differing forms of organizations allow for the development of women's initiative and leadership? Finally, how does organizational participation affect the attitudes and behaviour of women workers?

A recent study estimated that 64 per cent of the urban female labour force in India was in the IS (National Institute of Urban Affairs, 1987). Research on women IS workers in India has generated a wealth of case studies on IS women in particular cities, occupations and organizations (e.g. the studies contained in Singh and Kelles-Viitanen, 1987). This material has facilitated theorizing about gender in the Indian political economy in several recent papers (Bardhan, 1985; Kalpagam, 1986a, 1986b). Although the literature on IS women has provided valuable information and theoretical insights, it also has several limitations. In general, the studies are confined to the individual level of analysis. With certain exceptions, there is little information about the nature of the IS occupations in which women are found—their internal structure and functioning and how they fit into the political economy. Insufficient attention has been paid to the effects of policies and organizational forms among IS women.

Usually a sample survey methodology is used without adequate contextual and historical perspective. Often the research design is inadequate in addressing questions concerning the impact of IS work or organizational membership, as all those surveyed are workers or members. There is a need for multilevel analysis—examining IS woman in the context of their households as well as studying the structure of their occupations, the role of any organizations in the occupation and the impact of public policy. Such a multilevel analysis requires a combination of methods—survey research, organizational and policy analysis.

METHODOLOGY

Our study attempts to build on the decade of research reviewed on IS women in India and to overcome some of the limitations of the previous research. In light of the extreme heterogeneity that characterizes the IS, Trager (1985) suggests the study of specific IS occupations instead of the IS as a whole. We have selected five occupations in which women predominate and which are situated at different points along the several dimensions of IS activity sketched out in Table 1.

The heterogeneity of women's IS work in India can be conceptualized in terms of variation along a number of dimensions. These dimensions are similar to the different kinds of worker vulnerabilities outlined by Standing (1987). What sector or sectors is the work in? What is the labour status of the worker? What kind of enterprise does she work for? What gender roles exist in the work? To what extent are there caste barriers to entering this type of work? We thought that where women workers were located on these dimensions would influence the problems they faced and the types of organizations that might emerge in their occupations. Thus in our study we selected occupations located at different points along the dimensions of IS work, each representing large concentrations of women workers.

Two of the occupations involve family or self-employment. The fisherwomen combine trade and primary sector work (fish processing) and are involved in a family business with a strict

TABLE 1
Dimensions of Women's Informal Sector Work

	Subcontract	Sweepers	Fisherwomen	Khanawalis	Domestic
Sector	manufacturing	service	primary and trade	service	service
Labour Status of Worker	outworker	casual wage worker	family employment	self-employment	domestic worker
Nature of Enterprise	small-scale enterprise (dependent)	small-scale enterprise (independent)	micro-enterprise	micro-enterprise	private household
Gender Roles in Work	women only	men and women	gender segmentation	women only	men and women
Degree of Caste Homogeneity	multi-caste	Balmiki only	Koli only	multi-caste	multi-caste

division of labour by gender. It is the traditional occupation of the Koli community. The fisherwomen studied were from Versova, one of the original fishing villages of Bombay. The fisherfolk live in small houses down crowded lanes in Versova. The cooks or khanawalis offer eating facilities to male migrant workers whose families remain in the countryside. They practice a service occupation in their own homes and are helped by their daughters or daughters-in-law. The khanawalis tend to cook for men from their own *jati* and region and a majority are Maratha. They live in *chawls* (tenements) in the older working class areas—such as Prabhadevi—where the textile factories are located.

Two of the occupations involve casual wage work in personal services. Domestic service is a special type of casual wage work characterized by personalistic ties between employer and employee. It is a type of work open to all jatis, and it is a work that has become increasingly feminized in recent years. The women domestics we studied lived in Dharavi, a large slum between the Western and Central railways near Mahim. They each worked for several households in middle class colonies in Mahim and Matunga doing *chhuta kam* (part-time work) which involved washing dishes and clothes, cleaning the floors, and in some cases assisting in meal preparation. A majority of the women studied were Maratha.

In Indian households the tasks of cleaning the toilets and taking out the garbage are usually performed by sweepers. The sweepers in private buildings are casual wage workers who are almost always from jatis considered 'untouchable', who traditionally had been assigned these ritually defiling tasks within Hinduism. The group we studied were Balmiki from Haryana. The Balmiki jati forms part of the larger sweeper (*bhangi*) caste of North India; members accept Balmiki (Valmiki, the author of one version of the Hindu epic, Ramayana) as their patron saint (Das, 1982). The Balmiki community in Bombay lives in a number of slums along the Western Railway. Our respondents lived in slums in Bandra East and Mullund. Their homes were ramshackle huts, and the area resembled a village with chickens and pigs running about. While the domestic workers were employed by individual households, the sweepers were employed by the building society committees which managed the apartment buildings.

The last occupation—subcontracting—involves a type of casual wage work (outwork) in manufacturing. The term 'subcontracting' refers to a variety of work relations where a principal enterprise gets some of its work done with labour not directly under its own supervision (Nagaraj, 1984). Factories and other enterprises engage in a wide range of subcontracting operations, from giving work to other factories to giving out piece work to be done in homes. Subcontract workers do production work in their own home or in workshops. Both men and women do subcontracting, but women are concentrated in the lower levels working in homes and sheds in the slums. It is work open to all jatis and tends to be available near the industrial estates that ring the outskirts of Bombay such as Ghatkopar. The area we studied was a slum of huts climbing a hillside between Ghatkopar and Vikhroli. There men tended to work in factories and women tended to do subcontract work in their homes.

Thus we selected for study five IS occupations in which women predominated and with which organizations were associated. Our approach was contextual and comparative, comparing organizations and occupations in various localities in Bombay. Our study uses a variety of methods to build a picture of the women workers in the context of their household and occupation and as they are affected by organizational interventions, government policies and macroeconomic trends.

A survey of two hundred women—forty women in five occupations—was conducted in order to gain insight into the women's perspectives on their work, families, and organizational activities (see Appendix). The organizations themselves were studied through interviews with the (mostly male) leaders and activists, examination of organizational records, and review of newspaper articles. State policies affecting the occupations, questions of gender stratification in the occupations, and the effects of macroeconomic trends on the occupations were studied through interviews with researchers and government officials and through a review of the literature on the Bombay labour market.

The study places the survey findings within the occupational, organizational and public policy context in India through a comparative analysis of the five occupations. It examines stratification and segmentation in each of the occupations relating these patterns to women's roles in the occupations. It looks at the differential role of state policy across the five occupations. The study looks at the role of government policy in facilitating or impeding organizational formation and effectiveness in the IS occupations and at other factors shaping the emergence and the performance of the organizations.

In a study such as ours, issues of representativeness and sampling immediately come to mind. Because of the lack of information about the universe of women in the IS in Bombay, it is impossible to draw a 'representative sample' by conventional means. On the basis of secondary sources, census statistics, discussions with researchers and activists, and extensive forays around Bombay we have selected cases that maximize representativeness across several important variables: occupation (representing the most women and representing variation in IS work), organization (representing the types of existing organizations and representing variation in organizational characteristics), and locality (representing a range of lower class neighbourhoods in Bombay in terms of location, age, degree of heterogeneity of residents).

Although the five organizations associated with the occupations claimed to have membership lists, they lacked up-to-date membership rosters with addresses which we would have needed to select random samples of their members. The sampling procedure we used was quota sampling: we chose respondents across the range of ages available and tried to interview married, widowed, separated, and unmarried women in each occupation. (A description of the way the sample was selected and of the characteristics of the sample is provided in the Appendix). Our sample ended up being roughly representative of women in Bombay over 15 according to the 1981 Census in terms of age, marital status, and migration. Our sample was more heavily Hindu and less educated than the Bombay adult female population as a whole. (For a fuller account of our study, see Everett and Savara, 1989.)

ORGANIZATIONAL BACKGROUND

In what ways has occupational variation shaped organizational formation and development in these five cases? What have been the implications for women workers? Occupational characteristics have shaped the organizational forms that have develped. The casual wage workers in personal services formed unions. The fisherfolk in family-based businesses formed a co-operative.

Among the khanawalis and the subcontract workers the organizational form chosen was the society and trust. Although in one case the women were self-employed and in the other outworkers in manufacturing, both organizations in part resembled co-operatives: a credit co-operative for the khanawalis and an industrial co-operative for the subcontract workers. Thus the two main organizational forms in the IS occupations studied were unions and co-operatives.

There has been variation in the internal and external resources available for organizational development across the five occupations. The main external resources have been state policies and social movements. The state facilitated organizational development in some cases and impeded it in others. Social movements, political parties and non-governmental organizations provided inspiration and training to organizational leaders and inserted new issues and ways of looking at the world into the public agenda. The social and economic characteristics of the five occupations—labour status of the worker, gender roles, degree of caste homogeneity, type of economic enterprise, and economic sector have shaped the availability of internal resources, leadership, financial resources and unity. These occupational characteristics have been important in their own right and they have influenced the availability of external resources.

The differential state role in the five occupations has influenced organizational formation and development. According to Nirmala Banerjee (1988), Indian planners conceived of the unorganized (informal) sector as being made up of isolated self-employed workers instead of workers in a diversity of labour statuses. Moreover, the state did not treat all self-employed workers alike. Policy intervention in the fishing industry was initiated as early as the 1940s to promote modernization in order to increase production. Co-operatives were encouraged and funds were made available for boat mechanization, ice factories and other facilities through government credit programmes. In contrast the government did not make credit available to the self-employed poor in petty trade, production and services until the 1970s when a variety of 'weaker section' (a Government of India term for disadvantaged groups) lending programmes were developed by the banks after nationalization. Legislation to regulate the wages and working conditions of casual workers such as domestic workers and private building sweepers still does not exist. With the exception of *bidi* (hand-rolled cigarette) workers, state labour regulations do not cover home-based workers in manufacturing or many subcontracting operations either.

Why has the Indian state intervened in some IS occupations and not others? To some extent government economic development priorities have shaped patterns of intervention (Banerjee, 1988). After World War II and during the First Five Year Plan the priority of increasing food production led to a government focus on agriculture and fishing (National Planning Committee 1948; India, Planning Commission 1952). During the 1970s, government realization of the limited capacity of the formal sector to generate employment led to the use of credit programmes to promote income generation in the IS. No similar priorities existed for domestic workers or private building sweepers. Government intervention has also followed widespread political mobilization.

Social movements, political parties and non-governmental organizations—the nationalist movement, the Harijan movement, the Jana Sangh, the women's movement and other organizations—were important external resources in the five occupations studied. The way in which they were resources varied. In the three occupations in which men and women worked, a primarily internally based male leadership arose, inspired and/or assisted by social movements or political parties. In the two women-only occupations the barriers to internally based leadership were too great. The women's movement mobilized external individuals and associations to organize these women.

The Versova Fish Co-operative was formed in 1944 (and registered in 1947–8), by young Koli men who had participated in the nationalist movement and had learned organizational skills in jail.[1] At that time the Kolis were troubled by shortages and inflation, and were dependent on Muslim merchants for fishing supplies and trucks to transport the Koli women to market. Initially capital of Rs 1100 was raised by selling ten rupee shares, and the co-operative bought fishing supplies and sold them to members. The co-operative was also able to take advantage of government financing schemes for boat mechanization and transportations and later for an ice factory.

The origins of the domestic workers union, the Gharelu Kamgar Sangh (GKS) lay in the successful attempt by Ganpat Pandale and other servants in the same building in Bombay to gain sleeping and toilet facilities.[2] Previously, these workers had slept on stairways and had not been allowed to use toilets in the building. Shortly after this experience Pandale visited his village, and he talked to a social worker who pointed out that there were over 500,000 domestic workers in Bombay who needed assistance. The social worker gave Pandale the name of Raman Shah, a Bharatiya Mazdoor Sangh (BMS) leader in Bombay. The BMS was the trade union affiliated with the Jana Sangh (now Bharatiya Janata Party

(BJP). Together the two men formed the GKS in affiliation with the BMS. At that time the majority of members were men who worked as full-time servants.

The Safai Kamgar (sweeper worker) Union (SKU) grew out of numerous efforts to organize the sweeper community as a separate constituency and as part of a larger Harijan movement.[3] Harijans (Gandhi's term for the untouchable castes) protested against caste discrimination and sought government educational and economic assistance. Sweeper castes working for municipal governments demanded an end to demeaning aspects of their traditional occupation, such as carrying human excrement in baskets on their head. By the 1950s Bombay Municipality sweepers (garbage workers, street sweepers, and latrine cleaners) had achieved the status of permanent government employees with benefits and job security. The SKU was started in 1978 by Ramesh Bidlan, a Balmiki with a primary education who worked as a peon in a government office and who had been active in earlier organizations. In the course of his work with Balmikis, Bidlan encountered many private building sweepers with problems. His own wife did that work as well. So he started the SKU which was registered as a trade union in 1981.

The emergence of the women's movement in India during the 1970s changed the way many non-governmental organizations and activists looked at working class women. Prior to this, if working class women were noticed at all, they were the recipient of welfare approaches. The Anti-Price-Rise Movement in 1973 in Bombay demonstrated the militancy of women, as ten to twenty thousand marched down streets wielding rolling pins (Gandhi 1988a). The publication of *Towards Equality*, and the subsequent successful efforts of women leaders to include women's economic activities in government documents, contributed to an economic development oriented towards women (see Chakraborty, 1985). It was within this climate that Prema Purao, the former treasurer of the Anti Price-Rise Movement, began to organize the khanawalis into the Annapurna Mahila Mandal (AMM) in 1975.[4] She had been a trade union activist of the Communist Party of India (CPI) for many years, but as she said, 'I saw the problems of these women for the first time during the 42-day strike in the textile industry in 1973 . . . these women continued to feed the workers despite the fact that the workers had no income and thus could not pay the women for the food they ate' (cited in Savara, 1981: 51). In earlier years those with perspectives sympathetic to textile workers had tended to be critical of khanawalis, who they felt were overcharging the workers and feeding them low quality food.

Through talking to the khanawalis, Prema Purao learnt that their main grievance was the high rate of interest charged by grocers and money-lenders for the provisions the khanawalis needed for their work. The idea of approaching the banks for loans came from Purao's husband, who was a bank union leader. At that time the banks had special schemes to channel credit to the self-employed poor (see Everett and Savara, 1984), but few loans went to women. The khanawalis were unaware of the bank schemes, and the banks were unaware of the khanawali businesses. Working with the Bank of Baroda, Purao devised a system to enable the khanawalis to receive bank loans. The first group of loans (Rs. 1500 each) was disbursed to fourteen women in March 1976. AMM was registered as a society and a trust.

The emphasis on women's economic development promoted by the policy-oriented wing of the women's movement influenced a number of women's organizations and other non-governmental organizations to start 'income generation' schemes for women in the 1970s and the 1980s. When MAH/COOP, a social work agency, establised a foster children's project in Ghatkopar in 1979, it did not focus narrowly on the sponsored children but instead more broadly on community development through health, educational, environmental and economic programmes.[5] Its main economic programmes, designed to stabilize family incomes, were income generation schemes for women. MAH/COOP began the women's income generation projects in 1984 as independent small businesses producing school uniforms and school notebooks. However, as often has been the case with women's organizations and income generation, the inexperience of MAH/COOP in business matters led to financial losses, and MAH/COOP turned to subcontracting arrangements with large public and private sector firms.

Table 2 depicts the five occupations studied and the associated occupational organizations.

TABLE 2
OCCUPATIONS AND ORGANIZATIONS

Occupation	Name of Organization	Type of Organization
Fisherfolk	Versova Fish Co-operative	Co-operative Society
Khanawalis	Annapurna Mahila Mandal (AMM)	Women's Association
Sweepers	Safai Kamgar Union (SKU)	Union
Domestic Workers	Gharelu Kamgar Sangh (GKS)	Union
Subcontract Workers	MAH/COOP	Social Work Agency

ORGANIZATIONAL RESPONSES TO ECONOMIC PROBLEMS

To varying degrees the five groups of women experienced insecurity of income, difficult working conditions, and harassment from employers, customers, or merchants, but the types of vulnerability varied. The two groups of self-employed women had to worry about access to raw materials and marketing their products or services. The sweepers and domestic workers faced difficulties with their employers who were in superior positions in the social hierarchy. They could be fired or accused of theft without any recourse. The sweepers in addition faced harassment because of their caste status. Subcontract workers had no control over whether they would get work or not. What services have the organizations provided to their members? Have they contributed to any improvements in security of income, working conditions, rights vis a vis employers, customers etc?

The fish co-operative did provide needed services to the women fish sellers. The main activities of the Versova Fish Co-operative centred around economic assistance to members: financing for boats and engines, a co-operative store for fishing supplies, transportation services for the women fish sellers, and an ice factory and cold storage. The transport operation grew from one truck in 1947 to over twelve trucks in 1987 as well as five luxury buses which are used by the fisherwomen to take trips around India. The co-operative's ice factory provided ice to keep the fish fresh while the women travelled to market.

The main activity of the AMM was providing its members access to credit. During its early years the AMM primarily served as an intermediary between the banks and the khanawalis. The core of the AMM was the loan group: a group of women borrowers in a neighbourhood would act as guarantors for the loans and as a support group for each other. The groups would meet regularly to discuss problems with work or family, and for educational programmes. The organizational structure evolved into twenty local area centres each encompassing several loan groups. In each area the local groups chose two to three women from each group to serve on the area committee. Each area committee elected several members to a fifty to sixty member AMM Executive Committee.

AMM tailored the banks' procedures to fit the needs of its members and did some of the banks' work: investigating prospective borrowers, maintaining lists of borrowers and collecting payments. Working from Purao's home, the AMM collected fees from members to cover expenses and set aside a portion of the loan proceeds for a building fund in order to establish an AMM centre. The success of AMM in its early years was remarkable. Initially repayment rates were very high, ranging from 90 to 98 per cent for loans taken out in 1976-9. However, the textile strike of 1982-3 caused loan recovery problems and interrupted organizational growth.

The prolonged strike and subsequent reduction in the textile labour force made Purao realize it was necessary to develop some alternative income sources for AMM members. She also wanted to expand the organization's focus to address other problems faced by khanawalis through educational, health, shelter, and legal aid programmes. By obtaining funding from several foreign non-governmental organizations and from the Indian Government, AMM was able to open a multipurpose centre in Dadar in 1983 and a second centre in Prabhadevi in 1987. A catering section was set up in the Dadar centre to train khanawalis to cook meals and snacks for middle class office workers. In 1986 thirty-three women had completed training and twenty women were in training. Those trained were preparing tiffins (meals), filling orders for parties, and operating canteens at two government offices. A vocational training programme in tailoring was also set up. After training the women, AMM assisted them in getting bank loans for sewing machines and subcontract work from Air India for stitching napkins.

The establishment of a centre enabled AMM in 1986 to start the Annapurna Mahila Co-operative Credit Society, a co-operative bank which offered savings and loan facilities to its members. The centre was the site of other activities as well. In 1985-86 legal services were provided in over one hundred cases concerning domestic violence, housing, money lenders, and fights in the community. Approximately ten to fifteen women sought shelter at the centre for two to fifteen days each because of domestic violence during the 1984-86 period. Recreational and cultural programmes for members included weekly videos and educational tours of India every two years. Some planned services—such as health check-ups and a day-care centre—had to be discontinued because of lack of space.

The main problems facing casual service workers concern the power differential between them and their employers. Both GKS and SKU efforts centred around attempting to resolve disputes between members and their employers. Disputes ranged from accusations of theft to dissatisfaction with pay, demands for reinstatement, charges of harassment by employers, and disagreements over leave. The GKS formulated and distributed its own rules for employers and servants because the government failed to enact legislation regulating domestic service. This strategy, however, only worked in localities where the union was strong. GKS records suggest that it was intermittently active between 1972 and 1986. Its records show there were 202 complaints made by men and 102 complaints made by women during that period. Outcomes were listed for 156 complaints (108 by men and 48 by women) with about half of both men's and women's cases classified as successfully or partially successfully resolved.

The main activities of the SKU also involved handling the grievances of sweepers. According to union records, most of the complaints were by women (721 in comparison to 115 by men between 1979 and 1984). It appeared that most of the grievances were not successfully resolved. Bidlan and his associates attempted to negotiate with employers, held *morchas* (demonstrations) when negotiations failed, or took cases to labour court. The lack of legal protection for private sweepers put the SKU in a weak position with employers or at labour court. A leaflet put out by the organization described several cases in which Bidlan tried unsuccessfully to negotiate with housing societies on behalf of dismissed sweepers. In one case the secretary would not meet with Bidlan or respond to his letters, and in another the housing society replied to the SKU in English which the leaders could not read. The SKU had problems paying lawyers' fees, and its union registration was cancelled because the proper forms were not filed.

Both the SKU and the GKS also focused on lobbying for new legislation to regulate work in their respective occupations, but the GKS was much more active. The SKU brought out a leaflet in 1981 that argued for a law to regulate the working conditions of private sector sweepers, to ensure them status as permanent workers, and to guarantee them leave, medical and retirement benefits, but the organization did not appear to do much more than this. In part the greater efforts of the GKS reflected the weaker position of domestic workers in labour law. They did not fall within the industrial definition of a worker (because private households do not fall under the legal definition as an employer). Thus, no labour legislation applied to them, the GKS was not recognized as a union, and domestic workers could not bring grievances to labour court. However, the main reason for the two organizations' different levels of lobbying activity was the greater access to resources on the part of the GKS.

The GKS—in conjunction with other BMS unions—periodically organized demonstrations. In 1972 two demonstrations of 25,000 and 40,000 demanded legislative action, and in 1983 a demonstration was held to argue for the extension of legislation covering dock workers and head loaders to domestic workers (*Organiser* 5/1/83). In October 1986 several demonstrations were held to demand Diwali bonuses. In November 1987 a demonstration was held in front of the Labour Commissioner's Office to demand a minimum wage for domestics. On April 5, 1987 the BMS organized a demonstration of 35,000 to demand government recognition of the GKS and of domestic servants as workers as well as increased wages and bonuses (*Indian Post* 4/12/88). A BJP legislator brought this matter up in the legislature, and the Maharashtra Labour Minister replied that the government was studying ways to protect the interests of domestic servants (*Times of India* 4/16/88).

The new legislation regulating private sweepers and domestic service would potentially benefit women workers in these female-dominated occupations. However, organizations are necessary to translate any legislative guarantees into practical reality. The GKS, although relatively effective, seemed more attuned to the small minority (in Bombay the percentage is estimated to be 10 per cent) of male domestic workers. The SKU did focus on its female members, but it was relatively ineffective in resolving their work-related problems.

The main economic problem of women subcontract workers was the lack of steady income because work was not always available. MAH/COOP had hoped to provide a steady income source for women in Ghatkopar, but was not able to do so when it turned to subcontracting arrangements. In 1986–87 the projects included a stitching operation where women produced mosquito nets with velcro fastenings for one firm, conducted an envelope pasting operation for Western Railways, did napkin stitching for Air India, and ran a packaging operation for GKW, a multinational corporation with a factory in the Ghatkopar industrial estate. MAH/COOP officials claimed their goal was that these projects would eventually run on their own as separate co-operative societies or as one multipurpose co-operative.

There were approximately forty women working in the packaging operation. They sorted and boxed screws and also packaged various kinds of safety pins—bunching and weighing them, sealing bunches in plastic bags and labelling them, and putting the bags in boxes. MAH/COOP had been able to house the packaging operation and store the GKW materials without cost in a municipal building in the slum. The wages of a watchman and a supervisor were taken out of the earnings of the women workers. The women were paid on a piece rate basis, and in the first year only managed to get enough work for ten to fifteen days a month. Later there was enough work, but the women still made under Rs. 250 a month.

The MAH/COOP women workers experienced several types of problems working in the various income generation projects. The pay was low, the work irregular, and there were no benefits. When MAH/COOP attempted to enter the manufacturing business, inexperience led to mistakes in obtaining materials and in marketing the goods which meant MAH/COOP was unable to produce the products at prices competitive with similar products in the market. When MAH/COOP acted as a middleman for putting out work from large enterprises, there were difficulties in obtaining materials, disposing of finished products and receiving prompt payment. In many ways these income generation projects resembled the other types of work available to women in a slum near a large industrial estate: caning chairs for Godrej, rolling up plastic for kitchen scrubbers, and assembling electrical sockets. The main differences were that the MAH/COOP workers spent some or all of their work time in groups while the other women did the jobs entirely in their homes, and the MAH/COOP workers went to women's organization meetings while the others did not.

GENDER ISSUES AND WOMEN'S LEADERSHIP OPPORTUNITIES

The problems faced by the five groups of women workers related not only to a hierarchical economic system but also to the hierarchical gender relations that they experienced in their families, at work and in the community. To a greater or lesser extent they also suffered from caste hierarchy. Usually these multiple hierarchies were experienced together and not as separate forms of oppression. For example, a khanawali in debt to a grocer was forced to have sexual relations with him. However, in our survey, when asked what was the biggest problem they faced in their lives, the women most often mentioned lack of work for themselves or family members, alcoholism and domestic violence.

To what extent did the organizations address gender issues and offer leadership opportunities to their women members? Striking differences could be seen between the three mixed gender organizations and the two women's organizations.

The fish co-operative was run by an executive committee and officers elected by the members. There were no women on the executive committee. The Chairman, Mr Chikhale, maintained that women were not interested in leadership positions but that 100 per cent voted in co-operative elections. No gender issues have been raised in recent years by the co-operative. The practice of dowry has become common among the Kolis when earlier they had observed the custom of bride price. Although Mr Chikhale was personally critical of dowry, the co-operative did not take up this issue.

According to a recent study (School of Social Work, 1980), 90 per cent of Bombay's domestic workers are women. They predominate in *chhuta kam* (part-time work) which is currently the most common form of domestic service. The two GKS full-time organizers, the officers, and most of the activists were men. No gender issues have been raised by the organization. The leaders admitted that it was hard for them to approach women, and that women had no time to attend meetings because they had to do their own domestic work after they finished their paid domestic work.

From time to time the SKU was involved in attempts to promote social reform within the Balmiki community. In 1987 it distributed a leaflet entitled 'Stop Child Marriage and Dowry' brought out by the Balmiki Dalit Vikas Samiti (Balmiki oppressed development committee). The leaflet identified early marriage, dowry, the custom of women covering their faces in the community, and lack of education as social practices contributing to the backwardness of the community. A more popular activity of the SKU and other Balmiki organizations has been the annual celebration of Balmiki Jayanti (birthday), and efforts to get the day recognized as a national holiday. Balmiki leaders believed that an official holiday for their patron saint would improve the social status of their community.

The majority of disputes handled by the SKU concerned women workers, as women were numerically dominant among IS sweepers. To improve the community, leaders see many issues that need to be addressed to women directly for an impact. There are contradictions in practice, however. There were no women leaders in the organization. The community was very suspicious of outside intervention. Several years ago members of a Balmiki family contacted the Forum, a Bombay feminist organization, to report a rape of a family member by another Balmiki. When the Forum tried to investigate, they found that the individuals involved had been sent back to the village by the caste panchayat and no one in the community was allowed to talk to them (Gandhi 1988b).

The local AMM groups have provided a forum for raising gender issues to some extent. A description of a meeting in a recent AMM annual report (1986) illustrates the problems discussed. Members compared their charges for boarders and pointed out that without consistency in rates they would be divided against each other. They also talked about helping a member whose husband was beating her and the need to help any woman in crisis. AMM also offers services to khanawalis in trouble at its Dadar centre. Moreover, AMM has provided opportunities for leadership training for working class women who are the group leaders, executive committee and office bearers of the association. According to Purao, 'Economically what we have achieved is not a big thing for a bank, but for our women it is a revolution. The same women who were diffident about going to a bank are now running our credit society' (*Indian Express* 9/6/87).

MAH/COOP set up women's organizations (*mahila mandals*) which held regular meetings to discuss work issues, presented programmes on literacy, health, and nutrition, and offered yoga classes for the women workers. On International Women's Day there was a speaker on women's groups and songs on women's liberation. Primarily, however, the programmes embodied a welfare conception of women's needs. For example, experts taught the women how to prepare nutritious low-cost meals. The ostensible goal of the mahila mandals was to train the women workers how to run the operations themselves. Little progress seemed to have been made by 1987. A MAH/COOP staff member explained, 'Women to change their personality from housewife to entrepreneur will obviously take more time'. Even if the co-operatives are registered and the women workers assume the management role in the income generation projects, subcontracting arrangements do not offer much potential for improving wages and working conditions and for obtaining job benefits. The conditions of extreme labour surplus create a large power imbalance between the subcontract workers and the large enterprises.

ORGANIZATIONAL SUSTAINABILITY

In assessing the potential of IS organizations in improving the lives of women workers, it is important to look beyond current activities and examine how many people are being reached and what the future prospects of the organizations appear to be. In both size and sustainability the five organizations varied a great deal.

The smallest in size were the MAH/COOP's income generation projects—about 150 women worked in 1986–87, and the SKU, in which Bidlan estimated there were under five hundred members, mostly women, in 1987. Intermediate in size were the fish co-operative with 1179 members in 1987 (332 women who represented 28 per cent of the total membership) and GKS which its officials claimed had about 1000 dues paying members (60 per cent women). Largest was the AMM; a 1987 study by Acharya and Ramkrishnan estimated the active membership to be 6825.

The Versova Fish Co-operative has been remarkably successful in its objectives. Initially the leadership training acquired during the nationalist movement and sufficient financial resources within the community enabled the co-operative to start. Government schemes to assist the fishing industry have provided resources for continued economic improvement. Although the government schemes facilitated the development of the Versova fish co-operative, internal leadership and financial resources were crucial to the co-operative's success. In spite of government subsidies, many primary fish co-operatives operate at a loss: this was true for 58 per cent of fish co-operatives in India in 1972–73 and 40 per cent of fish co-operatives in Maharashtra in 1975–76 (Singh and Gupta, 1983). The success of the Koli enterprises in Versova appears due to the enormous local market in Bombay and the fact that Koli women (and not middlemen) control the marketing of fresh fish to Bombay consumers (Indian Institute of Management, 1981).

The existence of the 'weaker section' lending programmes of the nationalized banks facilitated the organizational development of AMM. In recent years, however, AMM has become disillusioned with the political use of bank lending through loan *melas* (fairs). The loan volume with the nationalized banks has decreased. With the opening of first one and then a second centre, AMM has become more centralized. The leaders spend more time at the centre and less time out in the community among the members. An administrative staff has been hired, and their education and middle class backgrounds separate them from the khanawalis. AMM has been successful in obtaining funding to support the retraining of khanawalis whose occupations face long-term decline. Yet the retraining programmes only reach a small fraction of the mass base of the organization. The leadership abilities of Prema Purao have been a central reason for the success of the AMM, but the excessive reliance of the organization on one person can be seen as a problem.

The two efforts at forming unions among IS women workers showed some differences and some similarities. The GKS mobilized the domestic workers for demonstrations, and the experience led to militancy on the part of many of the women. The party affiliation aided the union; the BJP contributed financial resources and turned out thousands of its other union members for domestic servant demonstrations. The SKU illustrated the difficulties of trying to start an unaffiliated organization in a disadvantaged community with few resources. It had not been successful in mobilizing women or in resolving grievances with building societies. The sweepers mainly participated in Balmiki birthday celebrations. Both the domestic workers' union and the sweepers' union shared a common problem: existing legislation did not give them a basis for protecting their members against employer abuse.

New legislation for domestics, and amendments in current legislation, are required to regulate working conditions in these occupations in order to facilitate effective organizational development.

The problems facing the SKU were overwhelming. The internally generated leadership lacked needed resources. There were three main leaders but no funds for paying full time workers. There was no regular payment of dues; people only came to the organization when they had a problem. The absence of legislation offered no standing for the organization in conflicts with employers. The leaders lacked the technical expertise necessary to participate effectively in the labour courts. They also lacked finances and connections to those in power. The potential membership was very poor and dispersed across 52 settlements. Balmiki organizations have been successful in obtaining municipal services for their community, but they have been generally unsuccessful in improving the working conditions of private sweepers or in reforming community practices. Demonstrations (morchas) in 1977 in front of ward offices resulted in the provision of water, electricity, and toilets by the municipality. In contrast, a 1964 demonstration demanding legislation for private sweepers led to the arrest of twenty-six men and three women.

The limited success of MAH/COOP in operating women's income generation projects is similar to the experience of other social work and women's organizations. They find it impossible to compete in the market as independent operators and turn to subcontracting arrangements with large enterprises. Although such arrangements offer little to the women workers, they are quite attractive to the large enterprises. For them, there is the public relations benefit of aiding charitable groups and promoting employment generation. Often, as in the case of MAH/COOP, a workplace, security and supervision are provided at no cost to the employer.

In attempting to protect their members from employer abuse, the domestic workers' and sweepers' organizations lack any legal resources. Forced to lobby for changes in existing policies, they are likely to turn to political parties for assistance. Organizations among the fisherfolk and khanawalis are able to base their organizations around existing government programmes to provide benefits and services to the IS. They face the easier task of ensuring the implementation of existing policies. Organizations such as MAH/COOP experience a number of problems setting up freestanding income generation projects for women, for there is a high level of competition in petty production ventures. Such organizations are pushed into subcontracting arrangements in order to sustain income generation projects. Under these conditions and the lack of any legislation on wages, the organizations are in weak position to guarantee regular employment or increased income for the women.

WOMEN'S PARTICIPATION AND PERSPECTIVES

The above analysis shows that to varying, limited degrees the five IS organizations were trying to improve women's access to material resources at work. However, the voices of the women are so far absent from our discussion. To what extent do they participate in the organizations? Why do they participate? How do they view the organizations? What do they see as the strengths and weaknesses of the organizations? How does organizational participation affect their attitudes and behaviour?

Levels of awareness, membership and activity in the organizations vary significantly across the five occupations. Almost all of the khanawalis, domestic workers and subcontract workers in our survey said that they were aware of and belonged to their organizations. While most of the fisherwomen said they were aware of their organization, only about half of those answering the question considered themselves members. Only twelve sweepers would give answers to these questions, and of these only one said she knew about the organization and was a member.

Thirty-three domestic workers, thirty-two khanawalis, twenty-three subcontract workers, ten sweepers and only one fisherwoman had participated in various organizational activities over the past year, as shown in Table 3. The variation in participation between groups suggests several things. It appears that the domestic workers and the khanawalis were the most involved in their organizations. While the fisherwomen knew about the co-operative, they themselves did not participate in it. The subcontract workers knew about and belonged to their organization, but only a little over half participated in organizational activities. The concept of membership in the SKU was very informal and the SKU was plagued by internal dissension. A quarter of the sweepers participated in it, but only one considered herself a member.

TABLE 3
NUMBER OF ORGANIZATIONAL ACTIVITIES BY OCCUPATION

No. of Activities	Sub	Sw	Fish	Khan	Domestic	Total
None	17	30	39	8	7	101
	(42.5)	(75)	(97.5)	(20)	(17.5)	(50.5)
One	16	6	1	8	12	43
	(40)	(15)	(2.5)	(20)	(30)	(21.5)
Two or three	7	4	–	24	21	56
	(17.5)	(10)		(60)	(52.5)	(28)
N	40	40	40	40	40	200

(column percentages in parentheses)

Three different indicators were used to measure the level of organizational participation of the women respondents: the number of different activities in which they participated over the past year, the types of activities, and how often they had participated in these activities over the past year.

The types of organizational activities they participated in varied across the occupations as shown in Table 4. The domestic workers had the highest incidence of participation in demonstrations, and the khanawalis had the highest incidence of meetings and of conferences with the leader. The types of activities most common among the sweepers and subcontract workers were attendance at cultural, social or educational programmes.

TABLE 4
TYPES OF ACTIVITIES BY OCCUPATION

Activities	Sub	Sw	Fish	Khan	Domestic	Total
Demonstration	–	1	1	7	30	39
Meeting	16	2	–	27	??	68
Met Leader	–	–	–	28	2	30
Cultural	15	14	–	8	–	37
N	31	17	1	70	55	174

(columns exceed 40 as some respondents had multiple activities)

TABLE 5
PARTICIPATION SCORES BY OCCUPATION

	Mean	Standard deviation
Overall Average	0.87	1.03
1. Subcontract	0.78	0.80
2. Sweeper	0.43	0.87
3. Fish	0.03	0.16
4. Khanawali	1.75	1.15
5. Domestic	1.38	0.81

A participation index was constructed by giving each respondent a score reflecting the number of different organizational activities in which she had participated—none, one, two, three—over the past year. Averages were computed for each occupational group as shown in Table 5. This table shows the high degree of average organizational activity of the khanawalis and domestic workers. Thus the average number of activities per woman ranged from a high of 1.75 for khanawalis to 0.03 for fish sellers. The khanawalis and domestic workers can be seen as the most active organizational participants, the subcontract workers and sweepers as less active organizational participants, and the fish sellers as non-participants.

The frequency of participation in the activities varied as shown in Table 6. The more active domestic workers (24) had participated in one demonstration in the last year and had attended meetings once a month. The more active khanawalis (30) met their group leader more than once a month and attended meetings at least once a month. The more active subcontract workers (13) attended meetings once a week and other programmes once a year. There were only two sweepers (and no fisherwomen) who participated in organizational activities more than twice a year.

TABLE 6
FREQUENCY OF ORGANIZATIONAL PARTICIPATION
BY OCCUPATION

Frequency of participation in the past year	Sub	Sw	Fish	Khan	Domestic	Total
Never	17	30	39	10	7	103
	(42.5)	(75)	(97.5)	(25)	(17.5)	(51.5)
Once or twice	10	8	1	–	9	28
	(25)	(20)	(2.5)	(0)	(22.5)	(14)
Three or more times	13	2	–	30	24	69
	(32.5)	(5)		(75)	(60)	(34.5)
N	40	40	40	40	40	200

(column percentages in parentheses)

We analysed the impact of factors other than occupation on the range of organizational activities and the frequency of participation of the respondents over the last year. In this analysis we excluded the fish sellers since only one had participated in any activities. The factors related to the extent of women's participation in organizational activities generally reflected the pattern of occupational differences: lower levels of activity among the sweepers and subcontract workers and higher levels of activity among khanawalis and domestic workers. As age increased, participation increased. Widowed/separated women and women with non-working husbands participated more than unmarried women or those with employed husbands. Women in households with fewer member were more likely to participate than those in larger households. There was no relationship between education or respondent's income and participation. As the per capita household income of respondents increased, there was a slight tendency for participation to increase.

Although approximately half of the women were active in their organizations and over a quarter exhibited high levels of mass level organizational participation, they were not involved in the leadership levels of their organizations. With the exception of one subcontract worker, nobody had ever held an office in their occupational organization.

There is an interesting pattern of reasons given for involvement in the various activities of the occupational organizations as shown in Table 7. The domestic workers participated in union activities to gain pay increases and for other family needs. Two domestic workers sought the intervention of their union leaders to get a place to live and to get school admission for their children. The khanawalis participated in their organization for work-related reasons as well: to get bank loans to provide working capital for their businesses. They also participated for other reasons: to go on trips around India with the group leader, to protest price rises, and to attend religious functions. The subcontract workers participated in their organization for work-related reasons and also said they wanted to learn about women's issues, health issues or other educational programmes.

Thus these two women's organizations were somewhat broader in focus than the domestic workers union. While the number of sweepers active in their organization were small, their reasons for involvement resembled both the women's organizations and the domestic workers' union. A major focus of participation was Balmiki day, the celebration of the birthday of Balmiki. Two sweepers were trying to build the sweepers' union and to participate in the caste panchayat. Three others sought the intervention of the organization leader to solve family problems in a manner similar to the interventions sought by the two domestic workers. In these cases the sweepers wanted housing, school admission and protection from *goondas* (thugs).

TABLE 7
REASONS GIVEN FOR PARTICIPATION IN ORGANIZATIONAL ACTIVITIES BY OCCUPATION

Reasons	Subcon	Sweeper	Khana	Domestic
Work Related	10	–	17	24
Pay Increase	–	–	–	27
Loan	–	–	36	–
Price Rise	–	–	5	–
Women's Issue	8	1	–	–
Religious or Recreational	8	7	8	–
Education or Health Programme	7	1	–	–
Organizational Development	–	2	–	–
Intervention of Leader	–	3	–	2
Total Number	33	14	66	53

(multiple reasons possible)

Relatively few women answered questions about problems

with their organizations or made suggestions for improvements. When asked if there were organizational problems, only 62 women replied and 53 of these stated that there were no problems. A further question asking what more the organization could do received only 27 replies, and seven of these were by subcontract workers who said that they had never thought about this question. Among the sweepers, khanawalis and domestic workers there were only three to six complaints and suggestions, and among the subcontract workers there were sixteen.

Even though these comments were few in number, they revealed common organizational weaknesses in all four occupations. Many of the comments were criticisms that the organizations had failed to solve work-related or other problems. Subcontract workers said they didn't get enough work and the organization should give more information on work. Domestic servants said their union should get them higher wages. One khanawali criticized AMM's loan procedures:

Money should not be taken for the building fund every time they give loan. Annapurna gives Rs. 2000 rupees as loan but we get only Rs. 1700. If we need Rs. 2000, then what's the use of getting Rs. 1700? We have to take Rs. 300 from others and give interest.

A sweeper said, 'The organization should do something about getting water and getting houses numbered'. A domestic servant wanted the union to help get some other work or to start a business. Another khanawali expected AMM to provide help for her children's education by giving clothes and books. These unmet expectations led some women to dismiss the organizations. A khanawali said of AMM, 'It will not do much'. A subcontract worker said, 'If there are problems, we solve them ourselves'.

Other criticisms centred around internal organizational problems. A sweeper said, 'The organization doesn't have enough money'. Some women felt that their views were not considered by the organization. A subcontract worker said, 'Even if we try to say anything in the meeting, nobody pays attention to us'. A domestic servant said, 'We think, but it's no use, so we don't say anything'. Another subcontract worker said, 'The organization is not like it should be. There should be unity in the organization and thinking should be the same'. Several women felt that their organizations did not keep in sufficient contact with members or hold frequent enough meetings.

One hundred eight women answered the question, 'What have been the effects of the organization on your life?' Seventy per cent of the responses were positive and thirty per cent said that the organization had made no difference. The responses by occupation are shown in Table 8. The subcontract workers and khanawalis were more likely to mention positive effects—experience, knowledge, unity, breaking isolation, problem solving, and getting a bonus—and the domestic workers were more likely to say that the organization had made no difference. One fisherwomen said that only men went to meetings of the co-operative, and another fisherwoman said she used to participate but no longer did.

TABLE 8
EFFECTS OF THE ORGANIZATIONS BY OCCUPATION

Effects	Sub-contract	Sweepers	Fish	Khanawali	Domestics	Total
Positive	29	4	2	27	15	76
No difference	7	—	2	4	19	32
Total	36	4	4	31	34	108

The positive effects mentioned included concrete benefits won. For example, several domestic workers discussed the bonuses received after they participated in a demonstration (morcha). One said, 'After the morcha we got one month's pay as bonus for Diwali. We have benefited a lot since the organization started'. Other women identified less tangible but nevertheless important gains. A khanawali said she came to AMM to get information about the world. Another khanawali stated, 'We have become independent since we joined the organization'. A domestic servant said, 'There is unity among women because of the organization'. For many of the women, organizational activities provided an important outlet. A subcontract worker explained, 'During meetings women can express their thinking and their problems'. A khanawali said that she liked to go to meetings, as she could get help for her problems. Another subcontract worker elaborated:

Living inside the house we don't understand anything about the outside world, how women are bound. They don't have courage to tell each other their problems. Through organization women come together, they tell each other their problems, and try to understand each other's difficulties.

As education increased, the tendency to answer that the organization had had a positive effect increased (gamma = .27). The more active a woman was in the organization, the more likely she was to mention positive effects: 56 per cent of inactive women mentioned positive effects, 67 per cent of the moderately active, and 81 per cent of highly active women also did so.

In what ways if any is organizational participation related to empowerment in the respondent's life in the household? We explored this question by looking at relationships between indicators of organizational participation and indicators of empowerment in family life: decision making scores and occupational aspirations for daughters (for more information on these indicators, see Everett and Savara, 1989). We found that generally the women who participated in organizational activities exhibited attitudes and reported behaviour in their family that were more empowering than those women who did not participate in organizational activities. In this analysis we excluded the fish sellers because only one fish seller had participated in organizational activities.

A decision making index was constructed from the respondents' answers to eight questions on decision making in the family. These questions pertained to son's and daughter's schooling, respondent's work, housework, food, loans, gifts, and religious expenses. Several variables—a woman's age, her position in the household, and her level of participation—were associated with variation in decision making scores. Older women and those women who were separated or widowed played the largest role in household decision making. As women's participation in the occupational organizations increased, their role in household decision making also increased.

The respondents were asked in what occupations they wanted their daughters to work. One hundred fifty-two respondents answered and mentioned seven different occupations as well as marriage. The occupations ranged from formal sector jobs such as service, nurse, teacher, and doctor to IS work such as stitching, homework and the respondent's occupation. We categorized the answers into three groups: formal sector, IS and marriage, and we analysed the data to see what factors accounted for the variation in occupational aspirations for daughters among our respondents.

Occupational differences, education, and participation levels were associated with variations in the aspirations the respondents had for their daughters. The sweepers were less likely than the other occupational groups to identify a formal sector job for their daughters: only 25 per cent of the sweepers mentioned a formal sector job in contrast to 47 per cent of the domestic workers, 58 per cent of the subcontract workers, 76 per cent of the khanawalis, (and 71 per cent of the fish sellers). As education increased, the tendency for the respondent to select a formal sector job for her daughter increased as well. Those who participated actively in their organization tended to have higher occupational aspirations for their daughters than those who did not (fish sellers excluded). As Table 9 shows, 37 per cent of the inactive women wanted their daughters to get a formal sector job, while this percentage increased to 73 per cent among the highly active women. As organizational activity increased, the tendency to identify marriage as the occupational aspiration for daughters also decreased.

TABLE 9
ORGANIZATIONAL ACTIVITIES BY OCCUPATIONAL ASPIRATIONS FOR DAUGHTERS

Aspirations for Daughter	Organizational Activities			
	None	One	Two or Three	Total
Formal Sector Job	16	16	29	61
	(37.2)	(47.1)	(72.5)	(52.1)
Informal Sector Job	5	7	6	18
	(11.6)	(20.6)	(15)	(15.4)
Marriage	22	11	5	38
	(51.2)	(32.4)	(12.5)	(32.5)
	43	34	40	117

(column percentages in parentheses)

This brief examination of the relationship between indicators of empowerment in the household and organizational participation suggests that those who participate in the occupational organizations in general demonstrate attitudes and behaviour that are more empowering than those who do not. We cannot prove that organizational participation 'causes' the increased empowerment in the household and work, but we have shown that there is a relationship between these two variables. What seems to make the most sense to us is that the same characteristics that enhance empowerment in the household also increase organizational participation. Thus it is the older women and the separated or widowed women who are more active. It is not surprising that these women also scored highest in household decision making. However, it is surprising and encouraging that these women had higher aspirations for their daughters than those women who were less active.

There was a high level of variation in organizational participation across the five occupations: the khanawalis and the domestic workers tended to be very active, the fish sellers tended to be inactive, and the subcontract workers and sweepers were in between. The nature of the organizations seems to account for at least some of the differences in organizational participation. The fish co-operative did not encourage women's participation. Annapurna Mahila Mandal offered a number of activities and practical benefits to encourage the participation of the khanawalis. The GKS had recently held demonstrations attended by many domestic workers. The subcontract workers' social work agency expected the subcontract workers to participate, but didn't address the workers' grievances. There were a few sweepers who were active, but the majority were estranged from the leadership of the union.

CONCLUSION

The common stereotypes of slum women as passive victims, mired in ignorance and tradition, are challenged by the women we studied. In spite of their long workdays, one half of the women were active in their occupational organizations and over one quarter were highly active. There was much variation in the levels of activity in each organization and in the effectiveness of the organizations in addressing problems faced by the women. The fisherwomen received some services from the co-operative and benefited from the boat engines their families had financed through the co-operative. They did not need to be active in the co-operative, which had evolved into an institution and was no longer a participatory body. The co-operative also did not encourage the participation of women. The fisherwomen were active in a city-wide fish sellers' association which in 1988 successfully rolled back municipal market fees after they had been raised.

The other group of self-employed women workers, the khanawalis, were not in an economic sector that had received four decades of government assistance. However, the AMM had been able to utilize the public sector banks' 'weaker section' lending programmes, started in the 1970s, as organization building mechanisms. AMM women were the most active of the five groups studied. AMM provided a forum for women to talk about their problems, the leader was able to influence policy at the national level, and the organization was able to acquire substantial resources (the centres and vehicles) from foreign funders and the Government of India.

Among the casual wage workers the domestic workers were relatively active and enjoyed some chance of obtaining the legislation demanded through their relationship with the BJP. In contrast the SKU had low activity levels, a high level of internal factionalism, and little success in negotiating with employers. The subcontract workers did participate in the MAH/COOP mahila mandal, but the organization could not always provide a steady income for the women.

To what extent have the IS organizations discussed here addressed obstacles to women's empowerment and material welfare? It should be noted that these IS organizations are relatively weak, often plagued by factionalism and serving constituencies subordinated by caste, class and gender. All the organizations provide some services to women and thus contribute some to their material welfare. Only the AMM provides women leadership training and an opportunity to raise gender issues. The success of AMM has depended on the availability of resources for organizational development: primarily the bank lending programme and to a lesser extent the financial assistance of foreign non-governmental organizations (NGOs). These types of resources have not been available in the cases of domestic workers and sweepers. The existence of established IS businesses, as in the case of the Kolis and the khanawalis, seems necessary in order to take advantage of resources available for organizational development.

Variation in economic sector and in the labour status of the worker accounted for variation in the availability of state policies to aid organizational development, as shown above. Variation in economic sector and in the type of enterprise in which the women worked shaped the amount of power the workers in Bombay could amass through forming collectivities. The fisherfolk with the large local market were able to get out of poverty through organizing a co-operative. Even if the MAH/COOP mahila mandal were to form an industrial co-operative for subcontract work, they would not be able to raise their earnings substantially. As a small-scale dependent enterprise, their co-operative would be constrained by the power of large enterprises in a labour surplus situation. The labour surplus situation affected the other three occupations as well. The khanawalis, domestic workers and sweepers had to prevent competition from others in their occupation.

Caste homogeneity appeared to facilitate organizational development. In the two caste-based organizations, leaders developed who were completely internal in contrast to externally based or at least externally supported leadership in the other three organizations. The fisherwomen and the sweepers faced less competition for work than the women in the three non-caste-based occupations, but they still faced some. Caste homogeneity enhanced organizational unity, but it wasn't enough to overcome the lack of financial and technical resources that plagued the SKU and contributed to factionalism in the Balmiki community. Caste homogeneity discouraged the raising of any gender issues except by community members themselves. This meant that the gender issues were raised by men and reflected a male perspective that may or may not have been shared by women.

There were severe limitations to what the organizations could do, and the nature and extent of these limitations varied across the five organizations. With the SKU its very survival was in doubt, as it lacked the resources necessary for organizational maintenance. The GKS activities appeared orchestrated by the BJP, and could be seen as being manipulated by party politics. The fish co-operative had evolved into an institution that efficiently provided services but had lost the sense of a participatory organization. The AMM was almost powerless against the downward trends in Bombay textile industry employment. In a similar manner, the fish co-operative was relatively powerless against growing water pollution that was beginning to threaten the fishing industry. MAH/ COOP was not able to establish independent economic enterprises and so failed to offer a real alternative to the subcontracting work available in the slums. Only a small number of women could be reached, which was also the case in the AMM retraining operations.

Even when the organizations were relatively successful in addressing the economic issues mentioned above, it did not necessarily follow that gender issues were also addressed. In the three organizations with male and female members, gender issues were essentially ignored and leadership opportunities were offered only to men. There was variation in the level of women's participation across these three organizations. There was virtually no participation by women in the fish co-operative, and a low level of participation by women in the sweepers' union. Only in the domestic workers' union, the GKS, was the level of women's participation high. The high level of participation was mainly due to the large demonstrations that the BJP-affiliated BMS assisted the GKS in holding. In the two women's organizations gender issues were raised to some extent and women had some opportunities for leadership. The participation of women was relatively high, and there were organizational spaces for the discussion of problems that the women faced in their lives.

With the exception of AMM's efforts to retrain a small segment of its members, none of the organizations were challenging the barriers which kept the women we studied in the jobs they held. All five groups of women were doing work which offered few possibilities of upward mobility and which was segmented by gender, sometimes increasingly so. The khanawalis were in the only all-women occupation. In addition to the traditional gender segmentation among the fisherfolk, the Koli men were going into other lines of work while for the women, fish selling and processing remained the main opportunities for earning income. Domestic service was becoming increasingly feminized. Among sweepers and subcontract workers there were patterns of stratification and segmentation in which women were concentrated in the lowest levels of the occupation. Thus among formal sector sweepers working for the Bombay Municipal Corporation, only one out of five were women, while among private building sweepers, a majority were women. Among subcontract workers, men were concentrated in small workshops and women were concentrated in homework.

The women in these five occupations were not seeking to overturn class and gender hierarchies through organizational participation, but they were seeking political visibility for their respective segments of the IS workforce. They were not on a level playing field: as women and as workers in occupations often uncounted in government statistics, their potential as a political force was uncertain relative to groups that had been organized for a long time and had greater economic resources and better connections to those in power. Even among the five groups the women faced unequal obstacles. The existence of the women's movement resulted in at least a nominal commitment on the part of government officials to women's participation in organizational decision making in state-recognized organizations. Thus even without a demand from fisherwomen, there was some effort among government officials to pressure fish co-operatives to include women on their boards. The domestic workers and the khanawalis were each pursuing different routes to attain political visibility: on the one hand, a party-affiliation strategy, and on the other, a non-party strategy. While AMM was relatively successful in its non-party strategy because of the existing government loan programmes, the existence of the women's movement, and the external funding it had received, the SKU with none of these resources was unsuccessful in its non-party strategy. MAH/COOP did not perceive of its mahila mandals as political organizations. In emphasizing only the income generation aspect of its organization, MAH/COOP effectively institutionalized the dependence of the subcontract workers on large enterprises which gave out work.

Do IS women benefit from participation in occupational organizations? A definitive answer to this question would require much more research, but the case study we conducted suggests that participation offers both opportunities and dangers for IS women. The opportunities and dangers exist at various levels. Within the organization women may have channels for expressing their views and gaining leadership experience or these channels may be closed to them. In the local context the organization may be either successful or unsuccessful in its objectives, so that women do or do not realize economic benefits through participation. In the wider political environment, the intended and unintended consequences of the organizations' activities may have implications for women's empowerment and material welfare over the longer term.

Organizations clearly offer some opportunities to IS women. Women's organizations may provide organizational space to solve problems and the chance for some women to develop leadership skills. Mixed or women's organizations with access to state policy resources and/or resources provided by external organizations may enable women to obtain tangible benefits through organizational participation. Through organization, IS women may gain a degree of political visibility which would make possible policy changes beneficial to them. Organizational participation may lead women to be more assertive in the household, workplace and community.

What are some of the dangers? In mixed organizations, women may be used for mass mobilization purposes without much chance for contributing their perspective or for attaining leadership positions. In some women's organizations the economic benefits and potential political influence of the organization may be extremely limited. In party-affiliated groups, women may only be part of the electoral strategy of the party and have no role in shaping party policy. In non-party organizations external funding may divert organizations away from their original objectives and lead to centralizing tendencies that neglect the mass base of the organization. Organizations that gain political visibility may be used to enhance the power of the leadership at the expense of the membership. Increased assertiveness in the household, workplace, and community may generate retaliation by those disturbed by a change in the status quo.

In July 1988, the National Commission on Self-Employed Women and Women in the Informal Sector presented its report, *Shram Shakti*, to the government. Its recommendations include expansion of existing labour legislation to cover various groups of IS workers and the creation of new procedures to facilitate organizations of women workers bringing complaints about wages and working conditions to the government. The organizational-facilitator approach embodied in Shram Shakti appears extremely useful for the groups of IS women workers examined in this paper. The problems facing unions of casual wage workers illustrate the need for state resources to address the needs of this sector of the workforce. In addition to advocating the extension of legal protection to IS workers, Shram Shakti calls on the government to support organizing efforts, including regional or central association(s) of workers, and to have IS organizations represented at the policy making and planning levels.

In the chapter, 'On Organising', Shram Shakti also argues that it is necessary for women workers to have separate organizations as well as being part of more general organizations that include men. Activist in rural grass roots movements have also made this point (Everett, 1986, 1989). Our case study has illustrated the need for separate organizations for women in mixed occupations, but the way to facilitate such organizations is not clear. Shram Shakti does not answer this question.

While organizational facilitation seems a crucial role for the state to play in order for IS women to increase their power and material welfare, it doesn't address all the economic problems facing IS women. For example, organizations of IS workers are unable to counteract many unfavourable macroeconomic trends affecting their occupations. Women in subcontracting, even if organized, are likely to be exploited. Thus the recommendations in the recent *National Perspective Plan for Women* (NPP) also should be adopted by the government. The NPP advocates expanded credit, other forms of economic assistance, and a National Training Institute for Women. These strategies offer some potential for increasing women's incomes and employment security through setting up co-operatives to market products independently and through retraining IS women for new occupations. The NPP recommendations should be made in conjunction with recommendations for state policies to facilitate the organization of IS women, not as a substitute for them. Only if organized will IS women have some chance to get their share of development resources.

APPENDIX

This research grew out of earlier research we had done on women IS workers and credit, and Mira's work with the AMM, which led us to want to study IS organizations and occupations in comparative perspective. The research design was developed through the writing of an approach paper and group discussions. The paper 'Preliminary Research Design: Women in the Informal Sector in Bombay' was discussed at a meeting at AMM in 1986 where activists and grass roots level workers involved with organizing women participated along with several women informal sector workers themselves. Based on these discussions, a preliminary household schedule was drawn up. This was translated and pretested in Hindi and Marathi. The final version of the questionnaire developed after two pretests on three of the occupations to be covered.

Our approach was to meet the organizations and do a interview with the leaders. We would give them a short booklet which described the research study. After getting the organization's approval, the organization's representative would take us to the site and make the field level contact with several members of the community. We would have a small group meeting with some women explaining what we were trying to do and why. Each woman interviewed was given a copy of the booklet. The field investigator would find out when would be a convenient time to come.

The researcher at the appointed time would meet the contact who would then take her to the woman's house to do an interview. After the interview the respondent would take her to the next person's house. As it worked out in practice the sample tended to be almost the entire community in a particular area that the organization said were its members. The interviews were carried out from October 1986 to July 1987.

We analysed the data using SPSSX to generate cross-tabulations, breakdowns, and Pearson's correlations. In some cases we list strength of association statistics (gamma and eta) in this paper.

The general characteristics of the women studied were as follows: 75 per cent of the respondents were married, 15 per cent were widowed or separated, and 10 per cent were unmarried. We sub-divided the married women according to the employment status of their husbands: 19 per cent of the sample had husbands in the formal sector, 41 per cent had husbands in the informal sector, and 14 per cent had non-working husbands. Their ages ranged from under 20 to over 70 with the majority falling between 25 and 44. Almost all of the women were Hindu (188), eight were Buddhist, three were Christian, and one was Muslim. Twenty-three castes were identified by the Hindu women with the largest caste group (76) being Maratha. Seventy-six per cent of the respondents were Maharashtrians, 20 per cent (the sweepers) were from Haryana, and the remaining 4 per cent were from other states. The majority (62 per cent) were born in villages and migrated to Bombay. Fifty-eight per cent had had no schooling. Collectively the women had had 727 children, 608 of whom were living at the time of the study.

The 200 women interviewed belonged to households with a total population of 1228 and an average household size of 6.14. The average per capita monthly income was Rs. 210, and ranged from a high of Rs. 523 for the fish sellers to a low of Rs. 157 for the sweepers. Clustered near the sweepers were the subcontract workers (Rs.165), domestic workers (Rs. 167), and khanawalis (Rs. 177). The respondents' average monthly incomes were as follows — fish sellers: Rs. 1186, sweepers: Rs. 321, domestic workers: Rs. 230, khanawalis: Rs. 215, and subcontract workers: Rs. 204. ($1 U.S. = Rs. 14). Fifteen per cent of the women lived in households in which they were the only worker, 49 per cent lived in two worker households, 18 per cent in three worker households, and 19 per cent in households with four or more workers. On an average the respondents contributed 42 per cent of the household income. There were 31 households with women earners only.

Fifty-eight per cent of the households lived in one room dwellings, 28 per cent of the households had two rooms, and 13 per cent had three or more rooms. Sixty-three per cent of the dwellings were *pucca* and 37 per cent were *katcha*. Only 7 per cent of the households had latrines in their houses, 55 per cent had latrines in their communities, and 38 per cent had no access to a latrine.

We are not that confident of the reliability of the income data, particularly of the fish sellers, and of the khanawalis, because the self-employed workers did not get a monthly wage, and they tended to be reluctant to share income information with the researchers.

NOTES

Earlier versions of this paper were presented at the Association for Asian Studies Meeting in Washington D.C. in March 1989 and at the Association for Women in Development Meeting in Washington D.C. in November 1989.

1. Information on the fish co-operative (Versova Machimar Vividh Karyakari Sahakari Society) is based on the organization's annual reports for selected years, membership files, interviews with the Chairman, Shri Chikhale, and other officers in Bombay during 1986–88, and Subbarao and Mathur (1981).
2. Information on the Gharelu Kamgar Sangh is based on organizational records and pamphlets, on interviews with Surakshan Mogde and other union officials in Bombay during 1986–88, and on newspaper articles (*Times of India* 4/16/88, *Indian Post* 4/12/88).
3. Information on the Safai Kamgar Union is based on interviews with Ramesh Bidlan and other leaders in Bombay during 1986–88, on organizational records and pamphlets, on Vivek (1986), and on Gandhi (1988b).
4. Information on the Annapurna Mahila Mandal is based on organizational pamphlets, annual reports, interviews with Prema Purao and other leaders in Bombay during 1981–88 and on Savara (1981), Acharya and Ramkrishnan (1987), Everett and Savara (1984), *The Daily* (9/13/87), *Indian Express* (9/6/87), and *Times of India* (9/4/87).
5. Information on MAH/COOP is based on interviews with staff members in Bombay during 1986–88, on organizational pamphlets and

on a telephone interview with GKW management in Bombay on July 5, 1988.

BIBLIOGRAPHY

B.T. Acharya and M.V. Ramakrishnan, *Annapurna Mahila Mandal: Study of Scope for Central Purchase and Distribution*, Bombay: Centre for Studies in Decentralized Industries, 1987.

Annapurna Mahila Mandal, *Report of Activities from October 1984 to September 1986*, 1986.

N. Banerjee, 'The Unorganized Sector and the Planner', in A.K. Bagchi, ed., *Economy, Society and Polity*, Calcutta: Oxford University Press, 1988, pp. 71–103.

K. Bardhan, 'Women: Work, Welfare and Status, Forces of Tradition and Change in India', *South Asia Bulletin*, 6(1): 3–16, 1986.

S. Chakraborty, ed., *Rural Women's Claim To Priority: A Policy Debate. Selected Documents from International and Indian Archives 1975–1985*, New Delhi: Centre for Women's Development Studies, 1985.

Committee on the Status of Women in India, *Towards Equality*, Ministry of Education and Social Welfare, Government of India, New Delhi, 1974.

B. Das, 'Untouchability, Scheduled Castes and Nation Building', *Social Action*, 32(3): 269–82, 1982.

Department of Women and Child Development, Ministry of Human Resource Development, Government of India, *National Perspective Plan for Women, 1988–2000 A.D.*, New Delhi: 1988.

J. Everett, 'We Were in the Forefront of the Fight: Feminist Theory and Practice in Indian Grass-Roots Movements', *South Asia Bulletin*, 6(1): 17–24, 1986.

——, 'Incorporation Versus Conflict: Lower Class Women, Collective Action, and the State in India', in S.E. Charlton, J. Everett, and K. Staudt, ed., *Women, Development and the State*, Albany: SUNY Press, 1989.

J. Everett and M. Savara, 'Bank Loans to the Poor in Bombay, Do Women Benefit?' *Signs: Journal of Women in Culture and Society*, 10(2): 272–90, 1984.

——, 'Women and Organizations in the Informal Sector in Bombay: A Study of Five Occupations', mimeo, available on request, 1989.

N. Gandhi, 'Masses of Women—But Where is the Movement? A Case Study of the Anti Price Rise Movement in Bombay, 1972–1975'. Bombay: mimeo, 1988a.

——, Personal Communication, 1988b.

C. Grown and J. Sebstad, 'Introduction: Toward a Wider Perspective on Women's Employment', *World Development*, 17(7): 937–51, 1989.

India, Planning Commission, *First Five Year Plan*, New Delhi, 1952.

Indian Institute of Management, *Marine Fish Marketing in India*. Volume IIA, VII. Ahmedabad, 1981.

U. Kalpagam, 'Gender in Economics: The Indian Experience', *Economic and Political Weekly*, 21(43): WS59–66, 1986a.

——, 'Women in Informal Sector "To Keep on Keeping on,"' *Economic and Political Weekly*, 21(51): 2216–18, 1986b.

R. Nagaraj, 'Sub-contracting in Indian Manufacturing Industries', *Economic and Political Weekly*, 19(31–33): 1435–53, 1984.

National Institute of Urban Affairs, 'Gender Bias in Employment of Women in the Urban Informal Sector', Research Study Series No. 20, 1987.

National Planning Committee, 'Report of the Subcommittee on Fisheries', in *Animal Husbandry, Dairying, Fisheries and Horticulture*, Bombay: Vora & Co, 1948.

A. Portes, and J. Walton, *Labour, Class and the International System*, New York: Academic Press, 1981.

M. Savara, 'Organising the Annapurna', *Women and the Informal Sector, Institute of Development Studies Bulletin*, 12(3): 48–53, 1981.

School of Social Work, *A National Socio-Economic Survey of Domestic Workers*, Mangalore, 1980.

G. Sen and C. Grown, *Development, Crises, and Alternative Visions: Third World Women's Perspectives*, New York: Monthly Review Press, 1987.

Shram Shakti: Report of the National Commission on Self Employed Women and Women in the Informal Sector, 1988.

Amarjeet Singh, and V.K. Gupta, 'Marketing of Marine Fish: Some

Policy Issues, in V.K. Srivastava and M. Dharma Reddy, ed., *Fisheries Development in India Some Aspects of Policy Management*, New Delhi: Concept Publishers, pp. 101–37, 1983.

A.M. Singh, and A. Kelles-Viitanen, eds., *Invisible Hands: Women in Home-Based Production*, New Delhi: Sage, 1987.

B. Singh, *Impact of Mechanisation of Fishing Boats in Maharashtra*, Bombay: Central Institute of Fisheries Education, 1976–78.

G. Standing, 'Vulnerable Groups in the Urban Labour Process', Working Paper No. 13, WEP Research Working Papers, Geneva: International Labour Organization, 1987.

P.V. Subbarao and Sondip K. Mathur, 'Parameters for Success of Versova Multipurpose Co-operative in Maharashtra', *Journal of the Indian Fisheries Association*, 10 & 11: 55–60, 1980–81.

L. Trager, 'The Informal Sector in West African Cities: Research and Directions', Paper presented to the African Studies Association, New Orleans, 1985.

P. Vivek, *Sweepers and Scavengers in Bombay: Professionals of the City*, Ph.D. Thesis, Department of Sociology, Bombay University, 1986.

10

Development from Within: Forms of Resistance to Development Processes Among Rural Bangladeshi Women

FLORENCE E. McCARTHY

INTRODUCTION

Among the many trends represented in the burgeoning women's literature are those focusing on the effects of development processes on gender inequality, and those analysing forms of women's resistance to oppression, exploitation and harassment. The literature on the costs of development for women began with Boserup's seminal work in 1970 and has since been the arena for extensive examination of how national and international processes of economic restructuring, industrial and agricultural transformation, and labour demand have affected all areas of women's lives. This area of research and policy analysis has also focused on specific forms of women's participation in and exclusion from institutions such as education, employment, health and politics. Given the breadth of information now available, it is apparent that women have been either ignored or only included in development processes as 'add ons' rather than as integral components of the planning and policy making process (Jiggins, 1984).

The literature on women and resistance incorporates diverse themes as resistance is a term which includes responses to nationalist struggles (Everett, 1981; Jayawardena, 1986; Omvedt, 1986) and class and economic issues (Bardhan, 1985; Liddle and Joshi, 1985). There is documentation as well of increasing forms of organized protest as women move directly into public visibility in defense of their rights and livelihood and against their exploitation and abuse (Kishwar and Vanita, 1984). On the Indian subcontinent, women's involvement in protest movements has a long, if largely undocumented, history, and is currently exemplified by the Chipko movement in India (Berreman, 1985) and the Women's Action Forum in Pakistan (Rouse, 1986).

This paper links issues of resistance and development by exploring the contradictory effects of the mobilization of rural women and the forms of resistance that are developing in response to development processes in Bangladesh. It is argued that the legacies of the 19th and early 20th century Indian social reform, nationalist and peasant movements contain implicit and explicit constructions of women which continue in the operations and assumptions of current development policies of the Bangladesh government and in local Bangladeshi non-governmental organizations (NGOs).

Constructions of women, i.e. ideas and assumptions about the 'nature' and capabilities of females, are also inherent in the plans and programmes of donor agencies as well. These formulations regarding the abilities and characteristics of women, and the 'appropriate' responses of males to them, converge in the plans and enactment of development programmes in Bangladesh as in other Third World settings.

Fundamentally, a gendered construction of women's roles in development activity assumes the limited, inferior nature of women and involves the creation of parameters of movement participation for females which mobilize their resources and creative energy without threatening established forms of partiarchal domination and control. Women are considered as dependent beings and their incorporation into development processes precludes provision for their acquiring access to formalized power or even to inclusion in collaborative decision making processes regarding policies or activities which directly affect them.

It is proposed that the processes of mobilizing women for development initiatives are contradictory in their effects and generate often unintended forms of resistance among women to specific programme activity as well as to forms of exploitation in the rural environment. The projection of women as passive creatures lends itself to a general lack of programmatic concern with the possible effects of programme participation on the women themselves, particularly changing forms of their own consciousness and willingness to assert themselves. It is assumed that women have no sensitivity or awareness of their situations and hence any deviation from the presumed docility of female programme participants is unexpected to programme staff.

Given the structure of domination and subordination which characterize women's lives, it is not surprising to consider that the active awareness of women of their condition remains largely unexpressed in either programmatic or familial settings. However, as the case suggests, one must theorize some form of consciousness informing the actions of rural women. The argument being presented here suggests that circumstances which provide rural women new opportunities for coming together, for increased social participation, and for new forms of experience contribute to new forms of collective action in some instances as well as to new (for them) interpretations and understanding of their social realities. Tapping these forms of consciousness is a difficult task yet is critical to the development of countervailing constructions of rural women which credit them with intelligence and agency.[1]

The following discussion will consider the development context in Bangladesh, clarify and define the term resistance, and explore causes of resistance among rural women. The determinants of resistance among women will be examined in terms of structures of domination such as the influence of familial and rural power relations which pattern forms of female subordination. It is suggested that resistance grows out of and in response to extent social conditions, particularly the intensification of personal distress in response to family impoverishment, loss of resources and limited employment opportunities. Resistance is also generated in a context where women increase their involvement in once forbidden social institutions. Participation in institutional relationships related to employment and schooling, for example, helps demystify the nature of social visibility and provides the shared experience of structured inequality inherent in such institutional arrangements. It is the confluence of these situations and experience that encourages women to become increasingly resistant to their conditions and become more active in their own defense. The paper provides three instances illustrating the forms of resistance in which rural women have been engaged. It ends with a brief consideration of the implications of such activity on our understanding of the complexity of rural transformation and its effects on the lives of women.

THE DEVELOPMENT CONTEXT IN BANGLADESH

The current level of foreign assistance to Bangladesh is approximately $ 1.6 billion a year. This assistance supports approximately 80 per cent of the Government's development budget as well as a portion of its revenue budget. In addition to providing food aid of about $ 380 million and commodity aid of $ 250 million in the period 1981 to 1985, project assistance was approximately $ 1,342 million which represents about 87 per cent of the total aid package (Third Five Year Plan, 1985). The ratio of debt service payments to export earnings increased from 13 per cent in 1979-80 to 21 per cent in 1984-85.

Funds allocated through project assistance extend into all

sectors of society. Increasingly, Government and foreign sponsored programmes are attracted to the work of non-govenmental organizations (NGOs) whose grass-roots approach recommends them as mediating channels for the mobilization into development programming of heretofore ignored segments of the population. The trend of foreign financing for NGO efforts is increasing as is the reliance on them for collaboration with government sponsored development efforts. This latter tendency is exemplified by a recent government request to twenty NGOs to ensure support and guidance to peasants awarded abandoned (*khas*) lands.

This request exemplifies the complex and contradictory nature of the development environment in Bangladesh. By asking NGOs to intercede on behalf of villagers against other rural interests also vying for control of khas lands, government is abrogating its responsibility to enforce and implement its own policies. At the same time, requesting NGOs to undertake such activities is likely to place them in direct conflict with established rural elites and jeopardize their ability to operate as relatively autonomous change agents in the rural areas.

In this context, NGOs may have limited options in fashioning ties to government if it is their clientele who are recipients of government resources such as tubewells, pumps, credit or land. The contradictory position of NGOs may ultimately threaten their long term utility to government as they may be deemed unreliable because of their acceptance of government resources in face of continued commitment to work toward realizing their own agendas. In the current political climate, many NGOs are caught in a double bind in that eschewing government linkages threatens continuing government recognition and licensing, while co-operating may entail loss of programme autonomy, changing status vis-à-vis rural people, and heightened conflict with rural elites.

The process of linking NGOs with government began in 1982 when government made it mandatory for all NGOs be registered and subject to periodic review. This included the regulation that all NGO funds were to be accounted for through government channels. The present situation finds more than 200 private development organizations and 18 major multilateral organizations operating in the country including the World Bank, the Asian Development Bank, and various agencies of the United Nations such as the Food and Agriculture Organization, UNICEF and the United Nations Development Programme. In addition, all the major bilateral donors such as the United States, Japan, Britain, the Netherlands, Sweden, Norway and the Danish government also provide development programme assistance.

The government's major development focus remains a production-population control strategy for continued economic growth. However, this strategy has been modified in light of the demands imposed by the IMF for economic restructuring. In order to maintain its Special Drawing Rights and general international creditworthiness, the Bangladesh Government has had to undertake a series of changes to bring the economy in line with IMF policies. This has required an increased emphasis on privatization of economic activity, devaluation of the currency, restrictions and modifications of trade and imports, and curtailed wage increases and support for labour unions and social services.

Additionally, in 1982 the Government undertook a New Industrial Policy (NIP) which encouraged industrial development through the denationalization of public enterprises, export oriented production, and provision of credit and other incentives to joint ventures and private entrepreneurs. Such policies are promoted in the attempt to stimulate employment, encourage domestic resource mobilization and foreign exchange, and reflect policy support for selected technological intervention (Third Five Year Plan, 1985).

The confluence of economic crisis with the change in political regimes and consequent change in State ideology emphasizing the Islamic character of the country, has marginalized even further the formal concern with women's issues. The government, in spite of varying degrees of public rhetoric, has devoted only a small portion of its resources, to 'the other half'. Government policy which in the late 1970s established a separate Women's Ministry, and devoted some attention to women's concerns, demoted the Women's Ministry to a Department and, in 1982, integrated it with the Ministry of Social Welfare. The Ministry of Women and Social Welfare was later merged with the Ministries of Youth, Culture, Mass Media and Sports. In the latest Five Year Plan, of a total budget of about 13 billion dollars, only 1 per cent will be allocated to the six combined ministries noted above.

Of the total government funds allocated to these ministries, women's affairs will receive about 16 per cent to be allocated for the construction of rural skill development and training centres, hostels for urban working women, providing day care, and maternal and child health. There is little in the plan indicating new or innovative approaches to women's issues or funding for such activity. Of the total figure of 2.4 million dollars allocated to women's affairs, $ 555,000 or 23 per cent is projected to come from the private sector primarily from NGOs (Third Five Year Plan, 1985). Other ministries such as Agriculture and Local Government and Rural Development are also expected to devote a small portion of their funds to women's activities.

The impact of the increasing Islamization of government rhetoric and activities on Bangladesh in general and women in particular has yet to be seen. The addition of the 8th amendment to the Constitution declaring Islam to be the state religion is only one indication of the increasing application of Islamic principles to state practice. There is concern among many Bangladeshis about the implications of Islamization for the legal and social status of minority populations, especially women, non-Muslims and non-practicing Muslims.

Additionally worrying is the utilization of Islam by the Ershad regime to justify and legitimate its own interests. The move to a more pronounced Islamic rhetoric represents a shift in national ideology and flies in the face of one of the critical issues dividing the wings of Pakistan which led eventually to war. The religious issue divided those in the West who assumed the superiority of their interpretations of Islam, against the representation of Islam in the East which was heavily overlaid with Bengali cultural influences (Roy, 1985; Eaton, 1984). The consideration by the West Pakistanis of the Bengalis as not 'true' Muslims was just one of many differences plaguing the state of Pakistan.

For many Bangladeshis, the change in the Constitution and the increasing Islamization of government policy represents a move away from a central tenet of the Bangladesh nation, that is the secular nature of the State and the guaranteed freedom of religion to all Bangladeshis.

THE NATURE AND TRADITION OF WOMEN'S ISSUES IN BANGLADESH

Basic to many of the current approaches dealing with women's issues are the traditions inherited from the Indian social reform and nationalist movements of the late 19th and early 20th century. These movements were noted for focusing on issues and mobilizing women in ways that appeared progressive without challenging the basic structure of male dominance and control (Sonalkar, 1984). For example, the reformist 'uplift' movement considered women's 'emancipation' as improving women's skills, knowledge and abilities related to their traditional roles as housewives and mothers (Karlekar, 1986). Implicit in this view was the assumption that women have little or no significant contribution to make to production, whether in the home or in the larger economy.

Also contributing to the development legacy is the influence of British, particularly Victorian, views of women which upheld middle class values of female education and limited social participation. These Victorian ideals were incorporated into colonial policy regarding the social position of Indian women (Kishwar, 1986). Women, in this view, were to reflect the status of their husbands and, therefore, education and limited degrees of freedom became necessary in creating wives capable of supporting the demands of rapidly industrializing economies (Tharakan, 1975).

Focusing on women's issues such as *sati* (wife burning) and the generally neglected nature of females was also a way the British justified their continued presence in India, as improving the status of Indian women became part of the 'White man's burden'. The sorry condition of women was used by the British as an indicator of the lack of readiness on the part of the Indians for self rule.

Policies favouring limited education for Indian women reinforced middle class interests and largely ignored the interests of peasants or working class people. Movements reflecting these concerns, however, have a long history and have contributed a legacy of more progressive development approaches in Bangladesh. Within most peasant movements and revolutionary parties active in Bengal during the rising nationalist tide in the late 19th and early 20th century, women played active roles. However, gender issues were generally considered secondary to issues of class struggle or working class demands. In some cases, such as the Tebhaga movement, women activists were aware of this discrepancy and chose to comply with the relegation of women's issues to a secondary position, assuming that in the long run the issues of concern to them would become issues for the general movement (Custers, 1986). In this, as in other movements, women were active partners but their issues were not taken seriously by male participants nor was their potential as activists fully utilized or recognized.

Translated into the current development scene, approaches to women's issues cluster around 'uplift' or social welfare traditions where organizations assume that external direction is required to improve the backward position of poor women. Poor women are treated as 'clients', and programme operations often encourage forms of dependency among participants for continued access to programmes for employment, credit, inputs and support. In these types of programmes women are not generally encouraged to exercise or develop personal skills through autonomous activity, or assume responsibility for or contribute to programme focus and direction.

These programmes, like many others, exercise a paternalistic, top down form of control within hierarchical structures where authority is not questioned. This approach to women may be found in local Bangladeshi organizations with or without foreign support, or foreign organizations running their own programmes in the country. In situations where staff are predominantly male, the degree of control and the limits to female member's autonomy is only enhanced. It is also the case, however, that class differences encourage middle class, educated women to assume the superiority of their experience and knowledge in constructing programme activities which they think are appropriate to meet the needs of poor women.

There is another equally strong approach which views women as pawns or players in the ongoing attempt to intensify resource mobilization and surplus extraction from the countryside. This is the view inherent in many government sponsored women's programmes. Many foreign donors contribute as well to this view of women. Women are looked upon as increasingly good credit risks, the key to reduced population growth rates, and critical to child health and family well-being.

Additionally, the conception of women as secondary income providers justifies their relegation to programme components which involve labour-intensive, low-capital technologies, or marginally productive, small-scale activities. Female programme participants are not generally viewed as human resources with individual and collective potential to be fostered and developed in their own right. Rather they appear to be viewed as vehicles by which national priorities can be realized.

Other approaches to women's programme attempt to encourage consciousness raising among female members, but within controlled activities such as literacy campaigns where alternative forms of action beyond organizational goals are rarely presented and where gender issues remain muted or unaddressed. Still other approaches stress grass roots organizing of both men and women, collective action, group control of agendas and decision-making, and shared responsibility for risks as well as benefits. Group solidarity is stressed within an organizational framework that supports varying degrees of decentralized authority to both staff and group members. Programme inputs and supports are in response to requests rather than top down allocations and staff are to facilitate rather than dictate group decisions.

However, even in these more progressive organizations gender differences appear in the treatment of men's groups vis-à-vis women's groups. For example, programme rhetoric may represent in its application only what affects or involves men's groups. Women's activities are both organizationally and actively separate (and never equal in the degree of concern or benefits directed to them). Programmatic inputs including credit tend to privilege male participants and groups. Loans to males, for example, even for the same activity, are larger than are loans to women (Sultana, 1988). This is the case, even though it is generally accepted that women are better credit risks than men. The organizational structure of most programmes finds males invariably holding supervisory positions over women and operating in paternalistic or chauvinistic ways in addressing women's concerns.

Some separate female organizations do exist which also follow the more decentralized, collective approach described above. In these cases, local rural women play enhanced roles in group activities, and power is more nearly equalized between staff and members. As will be elaborated upon later, it is among these groups that local women take the initiative in collective action against threatening social circumstances.

A DEFINITION AND CLARIFICATION OF RESISTANCE

Any attempt to reconstruct the history or image of women of the subcontinent has to confront the established ideology of them as passive, ignorant and dependent. This view of women is particularly pernicious in regard to Muslim females for whom the system of *purdah* is considered immutable and binding. What the recent literature on resistance suggests is that women have been much more active in the major social movements on the subcontinent than has been previously acknowledged. The question such findings raise is whether such activity was epiphenomenal and simply a response generated by the conditions of the time or indicative of the general characteristics or abilities of women.

The role of subaltern studies is to pursue similar questions in regard to forms of peasant history, culture and autonomy (Arnold, 1984; Guha, 1982). These writings challenge and attempt to rectify the historical record which has documented the actions, lives and events of the literate, the powerful and the wealthy. The focus of most traditional historical writing has ignored the majority of people, and hence our historical traditions suggest and reinforce the idea that except for brief, rare occurrences peasants, traders, craftspeople, artisans and small merchants were largely quiet and passive onlookers to the main events of the rich. What is argued by the subaltern scholars is that such groups had lives composed of action, ways of behaving and traditions, customs, and events, which where part of yet separate from recorded history. The recognition of those ignored in the selective construction of history acknowledges their agency however difficult it may be to 'correct' the historical record.

The social movements literature suggests that movements arise in response to persistent forms of dissatisfaction, inequity or exploitation that people experience. Movements are not epiphenomenal but are responses to contradictory relations in social environments. Conditions giving rise to movements can be understood through the analyses of social processes or conditions of a particular period. Implicit in this type of explanation is the proposition that social movement participation is an extension or elaboration of people's attempt to resolve, ameliorate or manage problems in their environments. Applying this interpretation of movement formation to women's activities in India suggests that female participation in social movements in the subcontinent during specific periods of unrest are not uncharacteristic behavioural practices but represent women's active involvement and responsiveness to their social conditions.

This interpretation of women offers a view of them that is active, responsible and responsive, and suggests that such a view requires exploration. A reconceptualization of women's responsiveness to life conditions as proactive rather than merely reactive, may contribute to an important revising of the stereotypic image of them as passive housewives and mothers uninvolved in productive activity and unable to defend their own and others' interests.

Essential also in any assessment of 'women' are the distinctions among them generated and influenced by conditions of class, caste, race, region and historical period. For example, while Bangladesh is racially fairly homogeneous, the treatment of Biharis and tribal groups suggest a strong racist tendency among the people (Bertocci, 1989).

Additionally, the patterning of women's participation in development programmes is class-related, as the very poor cannot afford the time nor have the resources to become programme members which requires the payment of fees, savings, and interest payments (Feldman, et.al., 1980; Sultana, 1988). The family resources upon which women can rely influences the nature of their programme participation. For example, women of more affluent rural families can rely on other female family members to watch children, feed husbands or share family work. Consequently these women are more likely to be given training options, or accept leadership positions in their villages.

Moreover, the degree of familial economic well being contributes to a woman's ability to obtain a loan, and to marshall the resources to make loan payments (McCarthy, 1990).

Regional differences are also important as they embody differential labour demand and employment opportunities. Even though Bangladesh is a small country, the difference in the salience of women's programmes as an income earning option is affected by regional variation in the agricultural/industrial productive base and the consequent demand for male as well as female labour. In Rangunia, for example, its proximity to the urban seaport of Chittagong with its attendant industries and small businesses, draws heavily on males from the region for urban industrial employment. As a result women have moved into labouring positions in agriculture, which pays comparably high wages to them. This is a disincentive to some women to participate in women's programmes.

Critical to understanding current forms of resistance is the realization that its modern definition implies forms of action which are overt and often violent. This is a view generic to struggles among those possessing power or those vying for access and control of it. Resistance, however, has its own place among the powerless and encompasses a range of behaviour of which overt action is only one alternative. Formal protest, letter writing, subversion, working to rule, studied ignorance, sabotage and pilfering are all forms of resistance by which dominated people respond in unequal power relationships (Scott, 1985; Turton, 1986). Resistance therefore refers to the myriad of ways that actors possessing differential power respond to their situations.

CAUSES OF RESISTANCE AMONG RURAL WOMEN

Forms of resistance among rural women can best be understood through an analysis of women's location within existing social formations, and the role played by family structure and rural power relations in restricting the options and freedom accorded them. Freedom refers to the social and legal rights and entitlements of women in addition to the flexibility and openness with which women can control and negotiate alterations in their activity, behaviour, social roles, decision making and resources including reproductive potential.

In Bangladesh, as in many parts of the subcontinent, women's subordination is embedded in family relations which enhance their low status through the young age of marriage, patrilocal residential arrangements, rigid age and gender hierarchies and the emphasis on circumscribed mobility and modesty codes stressing shyness, submissiveness and quiescent behaviour as indicators of family status. Personal status for women is dependent on being an uncomplaining hard worker and bearing many children, especially sons. While the actual specifics of any situation are greatly influenced by the class base and differential control and access of families to productive and social resources, in general the subordination of women is reinforced in legal and judicial processes by which women are denied or allowed only minimal rights to inheritance, divorce, or personal property.

Traditionally male dominance has been enforced by the control over women's physical mobility as well as their reproductive capacity. The physical seclusion of women within household boundaries has limited their interaction and experience with the outside world. The emphasis on childbearing particularly of males was and is a cardinal factor in the security and acquisition of status for women in their in-laws' or husband's household. Divorce or taking a second wife are frequent solutions to situations where women fail to provide children to a family. With the everdeepening economic crisis in Bangladesh, the marriage of daughters has become increasingly difficult as prospective grooms charge exorbitant dowries and dowry deaths and the commercialization of marriage contracts are becoming increasingly frequent.

In the countryside, social relations are structured by a highly differentiated set of social relations characterized by varying degrees of resource access and control. Increasing disparities in landownership, for example, result in approximately 11 per cent of rural households owning 52 per cent of the land while 68 per cent of households are landless or own less than half an acre (Januzzi and Peach, 1977). Elite control of land and productive resources establishes and reinforces their hegemony in the villages which incorporates tenurial relations, determinations of rent, sharecropping exchanges, wages and employment opportunities. Elite hegemony permeates social institutions such as the *salish* or local judicial bodies, the schools and mosques, and influences and constructs social opinion.

Elite control of land and production allows them to maximize their return from agriculture and provides them opportunities to diversify investments in enterprises such as trade, construction, cold storage, trucking or business. Social resources are also much more abundant among the elites then among other segments of the rural population. These resources are comprised of diversified social networks including government officers, townspeople, professionals and the police. Such resources are called upon to extend or promote local advantage or gain, to justify participation in government activities, to garner advantage in the distribution of inputs, and to provide protection and support in instances of legal difficulties such as overdue loans or corruption cases.

Patterns of domination and control also influence inter- as well as intrafamilial relations. For example, elite control of land and employment opportunities encourages competition among resource poor families for opportunities to work or cultivate land. As a result wages may be lowered, while costs of sharecropping or tenurial arrangements may rise. General insecurity among the poor increases as subsistence becomes increasingly difficult to secure. The response within resource poor families is the reallocation of family labour and the reliance on individualized forms of subsistence. This often results in the commitment of family members, including women, to wage based or income earning activity and often involves the reallocation of children's labour to household or child rearing activity.

Women of poor families become extemely vulnerable in these situations as they are unable to meet the dictates of purdah or seclusion and must work outside the household. This leaves them open to possible harassment or molestation from male villagers. As in India, there is little that male members of poor families can do to protect their womenfolk from sexual harassment by elite males.

Additional intrafamilial problems involve the increased burden on women to fulfill household reproductive and maintenance tasks in addition to working outside the homestead. In spite of employment and contributing to family income, many women are subject to abuse from husbands including beating, neglect, divorce and abandonment. In essence, women carry double burdens of insecurity as their vulnerability outside the homestead is replicated within the family in their relationships with husbands and other male household members. Male domination over women is shared by all men and becomes a powerful force linking males across classes in defence of their intrafamilial rights and prerogatives. Activities by and for women, therefore, are understandably often met with male hostility and disapproval.

Women's resistance grows out of contradictions existing within villages and families. For example, the numbers demanding work, as well as the institutionalization of contractual rather than patron-client relations, makes it difficult for village elites to control the activities of villagers who are not directly related or indebted to them. As a result, women may undertake innovative activities such as programme participation which elites may oppose but cannot stop. As a response to local Imams or elite males who oppose their activities women retort, 'I would be only too happy to stay home, but you will have to feed my children. Will you do this?' Or they respond, 'Will God put food in my mouth? If I don't work who will feed me?' (McCarthy, 1967; 1976). Faced with the reality of increasing numbers of poor women and an unwillingness to provide them with support, village elites desist in their opposition to women's activities.

Given severe economic constraints, family need supercedes male reluctance to have women leave the household or abandon traditional forms of purdah. In many cases, if costs of programme membership can be managed, it is eventually considered a better option for women than participation in the labour market. Women establish a means of benefiting from programmes or cease to participate in them. The contributions they make to family income may be direct, such as income derived from employment in programme centres, or may be indirect through the utilization of training, skills, credit or inputs acquired through programme inputs.

The negotiated participation of women in development programmes is related to a third contradiction which is that between benefits accruing to the family because of the improvement in the skills, abilities and knowledge of women, and male domination and control within the family. Women's experience outside the household establishes their own sources of knowledge and information and challenges the previously uncontested authority of male family members. This is especially the case in situations where women are able to demonstrate skills and knowledge in areas particularly defined as their own, such as gardening, poultry rearing, child care or home sanitation. Opposition to their learning is mitigated only by the contributions they can make to the family. As Bardhan points out, however, such activities do not necessarily result in an improved status for women within their households (Bardhan, 1985).

FORMS OF RESISTANCE BY RURAL WOMEN

The incorporation of women into development processes covers a wide range of activity from programme participation to wage dependent employment. In virtually every sector of activity women have experienced discrimination, harassment, unequal access to resources, fraud and systematic exploitation. This is the case whether receiving credit from the Grameen Bank, training from the Integrated Rural Development Programme (Feldman et.al., 1980), wage payments on rural public works programmes (McCarthy, 1983), or access to personal savings accounts (McCarthy, 1976).

The rhetoric of government regarding its concern for women has translated into few actual benefits for them. Among the most onerous situations for women has been the population control programme where the emphasis on sterilization has precluded education regarding other forms of contraception, and has subjected women to inhuman treatment in centres or mobile camps where tubectomies are performed in non-sterile conditions, with no physical examination, proper anaesthesia or follow up care being provided (Feldman, 1981; 1987).

The result of these experiences, in addition to those as wage dependent workers, has been to demystify the nature of social participation for once secluded women and confirm the finding that no part of the system represents womens' interests. Omvedt suggests a similar conclusion in her discussion of the development and changing nature of the women's liberation movement in India (1975; 1986). She argues that labour force participation with its attendant forms of discrimination, exploitation and harassment politicizes women and galvanizes them into action in defence of their own rights and interests. In India as in Bangladesh the adversity faced by women in social institutions generates an awareness of the lack of equity or opportunity present in existant social organizations and increasingly encourages women to organize to fight for their entitlements within programmes and institutions or organize outside of such organizations for their own interests.

In addition, the experience outside the home is often recapitulated within the family. The impoverishment of families through the loss of land and resources, the fragmentation of family networks and kin relations, and the increasing individuation of wage earning activity amidst limited, seasonal and low paying rural employment opportunities intensifies the personal distress and vulnerability of all family members, especially women. Women's low status and structured dependence on their husbands leaves them open to abuse and violence from family members. Often the equally impoverished condition of their father's household prevents women returning there in case of need. With the option for security and support diminishing within families, women are increasingly thrown back on their own initiative and personal resources for survival.

The forms of resistance in which women engage range from personal to organized activity. For example, some village women may secretly save money in case of personal need; others arrange share-cropping arrangements with women from more affluent families for poultry or small livestock (Feldman et al., 1987). Women also develop and maintain assistance networks in their neighbourhoods (*baris*) which include sharing work and providing child care and support in times of illness or emergency. Such ties are informal and loosely structured and often operate clandestinely.

The effect of mobilization for government or NGO projects such as public works or development activities often provides the context for legitimate association among women. While the original intent of such association may be programme related, they offer the context in which women discuss and exchange personal experience and generate mutual support for one another. It has been reported, for example, that women note the opportunity for friendship and support as a primary reason for membership in NGO or government programmes (McCarthy, 1967; McCarthy, 1976; Sultana, 1988).

What has also become apparent is the growing number of incidents where women are increasingly taking the initiative in organizing to protect their own interests. Illustrating this point are three case studies of programme member's utilization of their collective experience to respond to adverse social conditions. Without the experience of collective organizing, the development of group solidarity and shared perceptions of problems, and attendant changing levels of consciousness, these women would have been unlikely to respond as the examples show.

Case I

In the early Eighties, women's activities in the Chittagong district were being encouraged by *Nigeri a Kori*, an organization attempting to mobilize rural people around issues of group organization and collective economic activity. Nigeri a Kori provides little, if any, inputs to their member groups, but instead fosters group self reliance and group independence.

Women's groups had been formed in an earlier period and were basically self-directed in defining their own agendas and group activities. Group leadership and supervision was provided by local women, often those who themselves had begun as group members. The women's groups organized among poor, assetless women and attempted, through collective activities, to initiate savings, learning and income earning activities that promoted their own well-being and interests.

In one village, the group activities of poor women was viewed with dislike by certain landowning families, especially one influential village leader (*matbar*). He sent word to the women to disband their group and cease working in the village. His order was discussed by the women and they chose to ignore it. A further directive was sent to them and that too was disobeyed. As a next step, the matbar sent *gundas* (hired thugs) to threaten the women. The gundas went to the house of one of the group's leaders and began threatening her. As word spread of this occurrence, other women rushed to her assistance and the thugs were driven off. In response, the women went as a group to the house of the landlord to protest the actions of the gundas. While there, a fight broke out among the gundas of the household and the women. In the fray, one of the gundas lost an eye and the women dispersed to their homes.

As a response to the women's actions, and particularly with an injury which could be blamed on them, the landlord called for the police to come and arrest the group's leaders. Eventually, four armed policemen arrived in a jeep and proceeded to the house of a group leader. Upon hearing of the presence of the police in the village, the women banded together, attacked the policemen, unarmed them, removed their clothing and sent them out of the village on foot.

Not be to beated, the matbar called for the militia to intervene and they arrived in sufficient numbers to arrest numerous women and took them to jail. While the women had been dealing with the situation without the help or assistance of the parent organization, which had no supervisory staff living in the area, the arrest of their members led the village women to appeal to the Dhaka headquarters to intervene on their behalf. Their fear was that once in police custody, the women would be molested, injured and perhaps killed. Without funds, the remaining group members were unable to intervene on their members' behalf.

The appeal brought the Dhaka office into the picture, and through their connections pressure was brought to bear on the district police officers to leave the women alone. The Dhaka organization was also able to use family and other connections to raise sufficient money to obtain the release of the women on bail. Such intervention was an effective deterrent to villagers and local police, and the women's groups were largely ignored in the future.

Case II

In Faridpur District another women-led women's development programme operates. This programme operates with local women as leaders, focuses on poor village women, and stresses local group initiative in the generation and realization of group projects. The organization, which is known as *Shaptogram*, was initially started by an innovative, educated woman who decided to turn her energies to combating the problems of rural women. She chose her own village to begin her activities, and encouraged a type of organization that was self initiating and independent at all levels of organization. Women from surrounding villages as well as her own were encouraged to form groups, and general supervision and planning were the responsibility of local women as well. The object of the organization was to mobilize poor women, encourage their own skill and organizational development through a combination of activities stressing group planning, implementation and management of group projects, providing selected inputs and training, and encouraging group participation in developing, monitoring and changing the activities of the central organization.

Among the activities in which local women's groups chose to participate was road building work run under a government programme known as the Integrated Rural Works Programme. This programme was sponsored by Scandinavian assistance from Norway and Sweden and worked with the Public Works Department of the Ministry of Local Government and Rural Development. The IRWP programme was given responsibility for a selected number of infrastructural development projects throughout the country in which technical training as well as employment and income generating activities were to be promoted (McCarthy, 1983).

Poor, unemployed village people were hired to provide the basic labour on selected IRWP projects. In exchange for moving requisite amounts of earth each day, workers received wages or in-kind payments of wheat. Project wage rates were pegged to be roughly equivalent to average daily wage rates of male and female agricultural workers. The selection of small-scale, local projects was done by local government units known as Union Parishads (Councils); technical supervision was provided by the government engineering bureau and foreign consultants, and on-site supervision of projects was handled by a local project secretary appointed by the chairman of the Union Parishad. Union Parishads received payment for completion of their projects with the approval and presentation of full accounts to the local IRWP/Rural Works Bureau. Receipts included signed wage bills from the workers, validated pit measurements, accounts for inputs and materials and so on.

While ideal in theory, in operation the plan was marred by systematic cheating and exploitation of local workers, among them the Shaptogram groups. In one union in particular, the malfeasance of all union officials, including the Union Chairman and Project Secretary so enraged the women that they decided to protest. Gathering all the women and a number of male workers from a particular work site, the Shaptogram women marched to the local county seat (*Thanaro Upazila*) to protest their treatment to local government officials. Whether by chance or design, none of the appropriate Upazila officers were available when the demonstration arrived, and consequently the women decided to lodge a formal complaint with the local Deputy Commissioner, who is the chief government officer in the district. Arriving at his office the crowd of about 500 people encircled the building and would not leave until a promise was made that an investigation into the allegations of the women would be undertaken. The persistence of the Shaptogram women in pushing for their rights and fair wages led to the intervention of the District Martial Law Administrator. It was this person who was given the responsibility of holding a hearing at the work site in order to investigate the women's claims.

The hearing brought numerous government and military officials to the remote work site, as well as hundreds of project workers, local villagers and elected local government officials. During the course of the hearing it became apparent that the women were being cheated and those responsible for the project were seriously embarrassed by the testimony of the women. The District Martial Law Administrator chose not to pass any sentence at the time, but it was patently clear that serious cheating had been occurring.

The advent of the rains which washed away the pit measures and the delay of local officials to follow through on the decision of the District Martial Law Administrator to reimburse the workers meant that the women were never fully compensated for their labour. More dramatically, the Shaptogram district workers were blamed for the activity of the village groups and powerful Union Parishad chairmen attempted to destroy the organization by various acts of intimidation including threatening the lives of the chief workers, boycotting Shaptogram groups from obtaining further employment on road projects, and generally discrediting the organization. These activities were largely unsuccessful and over time a new form of employment has developed in which local groups contract directly with the IRWP/RWP for complete responsibility for road work. This appears to circumvent gross forms of cheating and provides a high standard of work by groups being awarded the contracts.

Case III

This case occurred within an organization in which the higher level positions dealing with programme planning, implementation and supervision are held by men and the primary objectives of the organization have been male-oriented. *Proshika*, as the organization is called, is innovative in its approach to rural people and has stressed local organization, initiative and planning, uses locally hired staff to assist group operations, and has promoted unique income generating activities such as contracting with local farmers for the sale and delivery of irrigation water required for local crop production. Low lift pumps, for example, are the property of a local group, and as a group, members enter into contracts with local cultivators for the delivery of water sufficient to raise their crops.

Among the activities more recently undertaken by Proshika have been those focusing on the mobilization of poor rural women. Similar to the formation of male groups, women's groups are encouraged to develop their own agendas and are supervised by programme workers who live in the local area. Proshika's experience with women is similar to that of other organizations, i.e. women are good credit risks, are serious about repaying loans and are responsible group members. The constraints of the rural economy limit what activities are possible for them to undertake, and these limitations are heightened by the reluctance of some male programme officers to push for innovative activities for women. Hence Proshika's women's groups follow activities similar to those promoted in many women's development programmes.

A striking difference, however, has been that among some groups, women have begun to define their own agendas separate and apart from those promoted by local leaders or programme workers. Of increasing importance among some Proshika women's groups have been issues of personal safety, particularly wife beating. Groups report that wife beating is quite common in their villages and at some point women began using their weekly meetings to discuss this and other village issues. In one village, the extensiveness of the problem and the women's negative reaction to such treatment led them to take action against a man who beat his wife who was a group member. The group decided on a unified approach, and warned the husband that if he beat his wife again, the group would discipline him. The warning was ignored, and when this woman was again beaten, the group retaliated by beating the husband. The women's group then let it be known that any villager who beat his wife would face similar treatment from them. The action of the women caused great stir in the village, and no formal action was taken against them by the village men. The women report that the number of wife beatings in the village has greatly reduced. Other Proshika groups report similar occurrences, and a number of other organizations in different parts of the country report similar responses of women jointly defending themselves from husbands' abuse (McCarthy, 1987).

The case studies differentially illustrate the argument made earlier regarding the conjoint influence of experience, conditions and relationships in generating and structuring forms of rural women's resistance. Particularly important are the patterns of domination which structure and set the parameters for women's action. All three cases involve poor rural women who are vulnerable to the actions of more powerful male family members, village leaders or government officials. Their vulnerability and lack of rights are inherent in their subordinate position in rural family and social life. In the past there has been no protection for women, other than that provided by the family, and intrafamilial violence or abuse could only be restrained through the intervention of other senior males or females.

In the first two cases, the assumed prerogatives of village elites towards the actions of village women were clear: women were to follow orders, and were to submit to robbery and systematic exploitation in the work environment. When women defied these orders or activities, the response was a general escalation of involvement of the social resources of the village elites as in the first case, or dismissal and later threatening behaviour in the second. The second case is interesting as it highlights an interesting situation in the rural areas. That is, men have very little experience with openly organized, defiant women, and hence their responses to how to respond to women are initially ill defined. During the hearing at the work site, for example, the District Martial Law Administrator, Deputy Commissioner, and various Upazila officials all unsuccessfully attempted to bring the vast numbers of women to order. They were totally ignored, and it was only as women began to think of their families and duties at home that they finally settled down and the hearing began.

The third case also illustrates the vulnerability of women within their own households, particularly to acts of violence by husbands. What is illustrated in the action of women is the effect of breaking the isolation in which women live and the resulting collective action to realign individual intra-family interactions.

The response of women in all three cases appears quite startling and unanticipated. Indeed, in some ways it is. However, it would be a mistake to assume that it was only the immediate events that caused women to openly defy the social order. It would also be mistaken to credit their organizations with creating the awareness among them which led to such action. The organizations are to be credited with bringing women together, and providing the opportunity for the sharing and collectivization of experience. However, as indicated, this aspect of mobilizing women is rarely given much attention by programme officials, and in all cases, the response of group women occurred without the direction or leadership of the formal programme hierarchy. Organizations are, at best, catalysts to a vast reservoir of rarely expressed knowledge and awareness held by women regarding their own conditions.

The combination of worsening socio-economic conditions, the limited resources, and lack of additional resources or options are posited as being instrumental in stimulating women to seek new forms of response to the conditions around them. But these conditions feed on, reinforce, or activate knowledge and awareness that women have acquired through their own life experiences. A number of comments made by village women in Faridpur illustrate how many women approached that conforntation with male authority. One women told the military and government officials, 'What do I have to lose in confronting you? My choices are to slowly starve to death or die quickly by a bullet'. Another women said, 'So put me in jail. At least there I will get three meals a day and have a *pucca* roof over my head' (McCarthy, 1983).

Superficial and simplistic constructions of rural women rob them of their agency in the literature as active and creative people. Theoretically what is amiss is that without exploring the complexity of women's awareness and consciousness of their life conditions, it is difficult to theorize the development of resistance as anything but a unique, random event. The paper argues that such is not the case, and that women's consciousness of their situation is profound, and informs their responses to structured domains of power and oppressive forms of subordination. It is the structure of the conditions of their lives that sets parameters to the forms of resistance women undertake. When situations arise in which new opportunities occur for bonding with other women, or developing countervailing social relations, women respond, and it is often these aspects of programme activity which in the long run are most important to them.

To summarize, this article has proposed that the constructions of women used in current development programmes implicitly assume their passivity, backwardness and ignorance. In fashioning activities for women, or in creating images of a 'typical' programme member, these assumptions filter into ongoing programme operations. What is absent from these formulations is a conception of women as active, creative, intelligent beings. In this regard, programme officials mirror images held of village women by much of the literature, as well as the members of the Bangladesh government and members of the donor community.

What the literature on resistance and the case studies suggest is the need to reconstruct the image of 'village woman' to account for what is a more likely and accurate depiction of them. That is, of women active, not just reactive; creative and innovative, not passive and ignorant; and courageous risk takers, not simple pawns. The point is not to romanticize the forgotten or ignored side to rural women, but to develop more accurate formulations of the complexity they represent. The implications for research and theory are immense, as the reconceptualization of who rural women are would lead to considerable changes in the understanding of gender relations, rural transformation, and certainly development planning and practice.

NOTES

1. Indicative of the distance between conceptualizations of rural women and their actual activity are assumptions about the nature of family and household existence. Given the poverty of most rural families, the question is rarely asked how such units survive, what keeps them together, or who manages to provide for the daily reproduction of the household. The literature will note the absence of males, changing household 'survival' strategies, and changing patterns of rural demographics which indicate that single person households tend to be female headed, not male. The problems of finding husbands for daughters is well known, as is the recognition of the pernicious nature of the dowry system. Yet women are never credited with managing familial responsibilities or being active in the pursuit of their children's or their own survival.

BIBLIOGRAPHY

David Arnold, 'Gramsci and the Peasant Subalternity in India', *The Journal of Peasant Studies*, 11(4): 55–174, July 1984,.

Kalpana Bardhan, 'Women's Work, Welfare and Status: Forms of Tradition and Change in India', *Economic and Political Weekly*, 15(51 & 56): WS 72–8, December 21–28, 1985.

Peter J. Bertocci, 'Resource Development and Ethnic Conflict in Bangladesh: The Case of the Chakmas in the Chittagong Hill Tracts', in Dhirendra Vajpeyi and Yogendra K. Malik, ed., *Religious and Ethnic Minority Politics in South Asia*, New Delhi: Manohar Press, 1989.

Gerald D. Berreman, 'CHIPKO: Nonviolent Direct Action to Save the Himalayas', *South Asia Bulletin*, 5(2 Fall): 8–13, 1985.

Peter Custers, 'Women's Role in the Tebhaga Movement', *Economic and Political Weekly*, 21(43): WS 97–104, October 25, 1986.

Richard M. Eaton, 'Islam in Bengal', in George Michell, ed., *The Islamic Heritage of Bangal*, Paris: United Nations Educational, Scientific and Cultural Organization, 1984.

Jana Everett, 'Approaches to the Woman Question in India: From Materialism to Mobilization', *Women's Studies International Quarterly*, vol.4. 1981, 169–78.

Shelley Feldman, 'Overpopulation as Crisis: Redirecting Health Care Services in Rural Bangladesh', *International Journal of Health Services*, (1), 113–31, 1987.

——, Farida Akhter and Fazila Banu, *An Analysis and Evaluation of the IRDP's Women's Programme in Population Planning and Rural Women's Cooperatives*, Dhaka: Bangladesh, 1980.

——, Farida Akhter and Fazila Banu, *An Assessment of the Government's Health and Family Planning Programme: A Case Study of Daukhandi Thana and North Mohammadpur and Charcharua Villages*, Dhaka: SIDA Development Cooperation Office, 1981.

——, Fazila Banu and Florence E. McCarthy, *The Role of Rural Bangladeshi Women in Livestock Production*, East Lansing: Michigan State University, Working Paper 149, 1987.

Government of the People's Republic of Bangladesh, *Third Five Year Plan*, Dhaka: Planning Commission, Ministry of Planning, 1985.

Ranajit Guha,' On Some Aspects of the Historiography of Colonial India', in Ranajit Guha ed., *Subaltern Studies VII, Writing on South Asian History and Society*, New Delhi: Oxford University Press, 1982.

F.T. Jannuzi and J.T. Peach, *Report on the Hierarch of Interests in Land in Bangladesh*, Dhaka: United States Agency for International Development, 1977.

Kumari Jayawardena, *Feminism and Nationalism in the Third World*, London: Zed Books Ltd., 1986.

Janice Jiggins, 'Rhetoric and Reality: Where Do Women in Agricultural Development Projects Stand Today?' Part I and 'Women in Agricultural Development: The World Bank's View', Part II, *Agricultural Administrative Review*, (3): 157–75, 1984.

Malavika Karlekar, 'Kadambini and the Bhadralok: Early Debates over Women's Education in Bengal', *Economic and Political Weekly*, 21(17): 29–31, April 1986.

M. Kishwar, 'Arya Samaj and Women's Education', *Economic and Political Weekly*, 21(17): 9–23, April 26, 1986.

M. Kishwar and R. Vanita, ed., *In Search of Answers: Indian Women's Voices From Manushi*, London: Zed Books, Ltd., 1984.

Joanna Liddle and Rama Joshi, 'Gender and Imperialism in British India', *Economic and Political Weekly*, 20(43): WS 72–8, October 26, 1985.

Florence E. McCarthy, *Bengali Village Women: Mediators between Tradition and Modernity*, Master's Thesis, East Lansing: Michigan State University, 1967.

——, interviews from 'Changing Family and Social Patterns among Bangladeshi Rural Women, Senior Fulbright Award Study', Comilla, Bangladesh, 1976.

——, *An Evaluation and Assessment of the Women's Component of the Intensive Rural Works Programme*, Dhaka: Swedish International Development Agency, 1983.

——, Field interviews with *Proshika* and *Basta Sheka* Women's Groups, Organizers, and Supervisory Staff, Dhaka: The Ford Foundation, 1987.

Gail Omvedt, 'Rural Origins of Women's Liberation in India', *Social Scientist*, 4(45): 40–54, November-December, 1975.

———, 'Peasants and Women, Challenge of Chandwad', *Economic and Political Weekly*, 21(48): 2085–6, November 29, 1986.

Shahnaz Rouse, 'Women's Movement in Pakistan: State, Class and Gender', *South Asia Bulletin*, 6(1): 30–7, Spring, 1986.

Asim Roy, 'The Social Factors in The Making of Bengali Islam', *South Asia*, (19): 23–33.

James C. Scott, *Weapons of the Weak: Everyday Forms of Peasant Resistance*, New Haven: Yale University Press, 1985.

Vandana Sonalkar, 'An Interpretation of the History of the Women's Movement in India from 1880 to the Present', in K. Murali Manohar, ed., *Women's Status and Development in India*, Warangal: Society for Women's Studies and Development, 222–30, 1984.

Monowar Sultana, 'Participation, Empowerment and Variation in Development Projects for Rural Bangladeshi Women', Ph.D. dissertation, Department of Sociology and Anthropology, Northeastern University, Boston, August, 1988.

Sophie M. Tharakan and Michael Tharakan, 'Status of Women in India: A Historical Perspective', *Social Scientist*, 4(4–5): 115–23, November-December, 1975.

Index

abortifacients, 46
abortion, 24, 31, 46
abuse,
    by employers 295
    by husbands 32, 39, 338
    family 340
    in the workplace 266
    resistance to 347–8
accumulation,
    and gender 127
    of capital 124, 146, 181, 205
    of labour power 120
    of land 130
activism,
    and women's issues 284
    of nationalist women 330
    of women IS workers 298–309
    women's 152–3, 250, 323
    and group protection 343
affinal village,
    husband's 75–6
    wife's 82–3, 85
age,
    and participation 301
    at cohabitation 93
    at marriage 36, 58, 72, 165, 247, 254
    voting 251
    of employment 264
age composition,
    EPZ work force 230
age distribution,
    effects of 157, 165
agency,
    and women 324, 349–50
agricultural labour,
    female 154–5, 191
agricultural management,
    women's role in 197–9
agricultural technology,
    benefits of 180
agriculture,
    and family labour 196
    share of employment 160
    in Bangladesh 220
AMM,
    and credit for women 286–7
    and women's activism 298–309
    founding of 284
Andhra Pradesh, 12
    dry zone geography 185–7
    employment in 156
animal care by women 30, 90, 105, 159, 199
armed forces,
    resources diverted to 267
assembly line,
    global 227
assets,
    and women's condition 167
    household 153
assumptions,
    about women 323
attached labour, 191

INDEX

authority,
   husband's 75
   compromised 339
autonomy,
   and docility 241
   and women's activity 330
   definition 27
   female 66, 68, 70–1, 111, 247
   for NGO programmes 325

Bandaranaike, S.,
   first woman prime minister 259–61
Bangladesh 215–42
   as an Islamic state 328–9
   development programmes 13
   NGOs in 323, 326
   war of independence 223
Bardhan, Kalpana,
   introduction 12
basic needs,
   in Kerala 162
Beneria and Sen,
   comments on 126–30
Bijnor, U.P., 11, 66–113
biological reductionism,
   danger of 69, 131
birth,
   rules surrounding 97–103
birth control,
   and the left 53
   methods 26, 27, 46
birth rates,
   decline in 254
birthplace,
   in Bijnor 97
Bombay,
   organizations 13
Bombay Presidency,
   demography 132–8
bonded labour,
   transformation of 191–2
breastfeeding, 107

brothers,
   collaboration among 87
Buddhism,
   in Sri Lanka 247
budget,
   Bangladesh's development 325

canal,
   Tungabhadra 185
canal building,
   and wage increases 203
canal villages,
   cropping patterns 185–6
capital,
   shortfall of 267
capitalism,
   and population growth 127
   logic of 118
capitalist agriculture,
   in Rajasthan 29
capitalist development,
   and gender 118, 148
   effects on women 3, 225, 262
   in agriculture 196
   policy for Sri Lanka 262
   politics of 252
   recency of 232
captive labour force,
   female 205
cash,
   male control over 200–1, 204
cash crops,
   introduction of 183
cash requirements,
   and women's work 196
caste,
   and agricultural zones 182
   and female leadership 260
   and gender 148, 310
   and the informal sector 277, 310
   and landlessness 154
   and women, studies of 5

discrimination 283
  in Andhra villages 186
casual labour,
  and patronage relations 192–4
casual workers, 280–1
categories,
  social construction of 7
cheap labour, 256
  women as 262
child mortality, 28, 71, 111
child rearing,
  middle-class 157
child survival, 129, 186, 331
childbearing,
  and human capital 67
  and women's status 28, 204
  conditions of 94–9
  decisions about 69
  mean age at 134
childbirth,
  experiences of 41–6
childcare,
  and the double day 123
  ideology of 118, 122
children,
  costs and benefits of 50, 54, 56, 111, 157
  investment in 147
  medical care of 83
  work in household 337
  work in agriculture 189
class,
  and demography 138
  and exploitation 196
  and fertility 15, 111, 129
  and gender 139, 146–9, 168
  and oppression 148
  and participation 334
  and population 131
  and reproduction 135
  and women's status 74, 128, 146, 336
  and women's work 198

class formation,
  according to Engels 124
  and domestic labour 123
  and gender 8, 120, 127, 149, 242
  and population growth 125
  studies of 5
  urban 218
clients,
  women as 330
co-operative,
  fishing 282, 286, 291–2, 294
co-operative bank,
  for women 287
co-operatives,
  women and 280, 290–3, 298
cohabitation,
  age at 93
colonialism,
  and demography 137
  and gender 3, 14
  and increased fertility 128
  and peasant economies 128–30
  and population growth 127–8
  and production relations 130
  and women's subordination 329
  and women's political role 249
commodity production,
  in the household 123, 297
comparative advantage,
  in cheap labour 228
comparative studies,
  gender in 9
competition,
  and IS women 310
confrontation,
  with male authority 349
conscientization, 13
  in IS organizations 304–8
  unintended 348–50
consciousness,
  and gender 7–8

and political action 13
and rural women 324
historical analyses of 8
lack of 124
of organizations 297
construction,
　women in 262
contraception,
　and sexual relations 60
　knowledge of 46
　lack of access to 340
　use of in Sri Lanka 255
contradictions,
　and gender 117
　and inequality 215
　between class and gender 129
　between culture and structure 224–5
　between education and skills 239
　between skills and status 339
　between women and households 55
　demographic 129
　in mobilizing women 324
　in villages and families 338
　labour market 229–30
contributions,
　made by women 219
　to family income 338–9
control,
　by rural social networks 337
　by the state 218
　ideological 158
　male 234, 336
　networks of 232–3
　of property 124
　over household finances 200–1
　patriarchal 184
　sexual 40, 135, 150
　urban labour market 241
　women's lack of 110

cooking,
　women's responsibility for 86–7, 104, 106, 218
cooks,
　in Bombay 277
　vocational training for 287
credit,
　for poor women 284, 286
credit,
　from Grameen Bank 339
　programmes for women 346
credit co-operative,
　for women 287
credit institutions,
　women's 226
credit relations,
　agricultural 199, 203
cropping patterns,
　effect of 179–80
　in canal villages 185–6
cross-cultural studies,
　women in 3, 4
cultural activities,
　women's participation in 250
cultural criticism,
　and gender 7–8
culture,
　and political character 247
　and regional demography 138
　and structural change 218
　and women's education 254

daughter-in-law,
　role of 83–8, 92, 95
daughters,
　aspirations for 306–8
　desire for 45
　difficulty of marrying 336
　education of 237–9
　production of 134
decision-making,
　women excluded from 202, 249

women's participation in 306–7
deconstruction,
  and gender relations 7–9
deference,
  female 82
defilement,
  at time of delivery 98–9, 106
delivery,
  recovery after 44–5
democracy,
  women's role in 257
demographic history,
  of India 115
demography,
  and colonialism 137
  and gender 14
  and patriarchy 158
  and women's work 121
  class dynamics and 131
  Indian 132–8
  of EPZ employment 240
demonstration,
  organized by women 345
  by IS unions 289
demystification,
  for rural women 340
dependency,
  agricultural labourers' 196
  ideology of 332
  of women on relatives 234
  on development agencies 330
  regarding cash needs 201
  women's 94, 182, 195
development programmes,
  and patriarchy 323–4
development strategy,
  capitalist 55, 154
  export-led 227–8
Dhaka,
  employment in 224
dialectic,
  demographic 129, 131

differences,
  structure of 146–53
discourse analysis,
  and gender ideology 118
distribution,
  unequal 204
division of labour,
  domestic 215
  gender and 219–20, 277
  global 217
  household 161
  sexual 118–19, 123, 158, 180, 182, 190
domains,
  male and female 219
domestic labour, 26
  and class formation 122–3
  and subsistence production 218–19
domestic servants,
  protection of 289
domestic service, 263, 277
  legislation 288–9
domestic sphere,
  and female autonomy 73
  conceptually isolated 67–8
  politics of 85
  relations in 215
  Sri Lankan women in 248
  women's role in 200
  women's work in 218
domestic worker's union, 282
  women's participation 301–3
domination,
  and appropriation 147
  resistance against 347
donor agencies,
  assumptions about women 323
double day,
  and IS workers 292
  burden of 338
dowry, 72, 74, 81, 292
  and murder 335

INDEX

drinking,
    husbands' 32, 201–2
dry agricultural zones 180–1
dualism,
    between public and private 9, 67, 121, 139
dung-work,
    women's responsibility for 106
Dyson and Moore,
    comment on 70–3, 137

ecology,
    historical 183
economic decline,
    and politics in Sri Lanka 257
economic development,
    for women 284
educated women,
    and development programmes 331
    as cheap labour 265–6
    employment in EPZs 229
    in Bangladesh 222–3
education,
    access to 254
    and class 8
    and employment 14, 15
    and hierarchy 150
    and organizations' effects 305
    and urban employment 237
    female lack of 71–2
    for poor girls 161–2
    incidence of 163
    of women in Sri Lanka 246–7
    of women in Bangladesh 237–9
    universal 253
    women's 156–7, 249, 330
electoral process,
    women in 257–8
electoral strategy,
    and IS organizations 313
elites,
    and landholding 336–7
    and vote competition 251
    and women in politics 258–9
    Bangladesh rural 325
    Bangladeshi urban 228
    in Andhra villages 197–8
    weakening control of 338
Emergency,
    and population policy 27
emic view, 8
employment,
    accessibility 156–7
    and kinship obligations 233
    female 255
    female, in Bangladesh 221
    generation of 226
    in canal building 185
    in EPZs 215
    in road building 344
    landholders' control over 337
    non-farm 221
    of female labour 191
    professional 224
    prospects for 255
    rural 226, 340
empowerment,
    of women by organizations 305–8
    through development programmes 339
    unintended 313
Engels, 124
entrepreneurs,
    and new cash crops 186
    Muslim 232–3
    new urban 241
epidemiology,
    and regional demography 138
EPZ employment,
    women in 229
Ershad regime,
    and Islamic state 328–9
ethnic conflict,
    in Sri Lanka 267

ethnicity,
    and gender 105, 110, 148
    role of 248
etic view 9
Everett and Savara 14
excess female mortality,
    in Bombay Presidency 136
    in India 131
    in U.P. 138
exploitation,
    and IS workers 315
    and patriarchy 205
    by class and gender 196
    defiance of 348
    of women 146, 225, 324
    through cheating 344
    women's resistance to 323
export,
    manufacturing for 224
export processing zones, 13, 164, 215
    establishment of 223
export production,
    effects for women 229
export-led strategy,
    and war situation 270
    in Bangladesh 327
extramarital sex,
    in village life 36

family,
    stresses on 266
    women's roles in 256, 335–6
family connections,
    importance of 248
    and women in office 259–60
family labour,
    in agriculture 196
    reallocation of 337
family planning, 26, 46, 61, 128, 254
family resources,
    male access to 253

family size,
    increased 254
family status,
    and female urban employment 238
    and women's behaviour 224
family survival,
    middle-class strategies 237
family wage,
    decline of 240
famine relief,
    and canal building 185
Feldman, Shelley 12
female employment,
    constraints on 218
female labour,
    by cropping pattern 184
    demand for 217, 240
female labour force,
    in informal sector 274
female networks,
    for neighbourhood assistance 341
female schooling,
    and employment access 164
female sexuality,
    control of 35
female subordination,
    to husbands 72
female survival,
    Indian 131–6
feminist critique,
    of social order 269
    of export processing zones 164
feminist movement,
    growth of 268
feminist scholarship, 2–5, 146
feminist strategies,
    Indian 168–70
feminist theory 116
fertility,
    and child mortality 28, 71
    and class 129

and colonialism 128
and gender 131
and poverty 54
and social relations 14
control of 111
declining 254
effect of employment on 165
in Africa 129
Indian marital 142
of peasant households 166–7
fighting,
    among women 89
    between women and *gundas* 342
financial intensification,
    and women's workload 203
first birth,
    age at 165
    lack of preparation for 42–5
fisherwomen,
    union's benefits for 286
fishing,
    modernization policy 281
    threatened by pollution 311
fodder collection,
    by women 30, 95–6, 161
foreign aid,
    to Bangladesh 325
foreign capital,
    opening Sri Lanka to 262
foreign exchange,
    need for 227
Foucault,
    influence of 7
fragmentation,
    of family networks 340
franchise,
    male 249
free trade zones,
    women in 262
    employment rates in 263
freedom,
    women's levels of 335

functionalism,
    and social reproduction 125

garment production,
    in EPZs 231
gender,
    and caste 310
    and class 139, 146–9, 179
    and class formation 242
    and contradictions 117
    and cultural criticism 1, 7–8
    and demography 116, 138
    and mortality 131
    and multiple hierarchies 291
    and poverty 156
    and resources 224
    and systems 10
    as a relation of production 117, 140
    as a marginalized category 5, 6, 7,
    in material context 1, 3
    in world perspective 3
    social construction of 182
    theory 3, 4
gender issues,
    and IS organizations 292–3, 311
gender norms,
    and female employment 165
    transformation of 158
gender relations,
    cultural construction of 216
generational growth,
    Indian 134, 142
generational mobility,
    and women 130
    of labour relations 242
    of occupations 237
    of property 125
    of women's occupations 306
gift-giving,
    by in-laws 82, 85

global economy,
    women in the 3, 4
global manufacturing,
    Bangladesh's role in 228
government,
    women's employment in 222
government ministries,
    concerned with women 328
government policy,
    and assumptions about women 323
    and industrial growth 226
    and IS workers 279, 281
    and population 26
    recommendations for 314–5
    regarding NGOs 325–6
Grameen Bank, 226, 339
Green Revolution, 12, 110, 179, 185
    effects on families 93
    Jats and 29
gross reproduction rates,
    Indian 135, 142
Gujarat, 128

handicrafts,
    women's work in 226
harassment,
    of working women 230, 285–6, 337
health,
    and pregnancy 28
    class effects 31
    free provision of care 252
    women's 15, 32, 100–1, 107, 136, 160–1, 165
    women's interest in 250
hegemonic control,
    state's role in 218
hegemony,
    and gender ideology 225
    of landholding elites 336–7
hierarchy,
    and education 150
    and government policy 331
    and population growth 128
    and skill assets 168
    and women's awareness 348–50
    caste and ethnic 150
    in reproduction 118
    in the family 85, 88–9, 148–9, 151, 181, 291
    of age and gender 336
Hindus,
    in Bijnor 70
historical demography,
    and colonialism 137
historical perspective,
    and systems 10, 14
    on Sri Lankan women 248
historical theory, 130
history,
    and social reproduction 125–7, 139–40
    rewriting 333
household arena,
    and reproduction 25
household composition,
    and women's activism 301
    effects of 67, 102
household consumption,
    storage for 203
household labour,
    and traditional relations 192
    reproduction of 216
household production, 123
household reproduction,
    women's responsibility for 200, 338
household strategy,
    income diversification 240
household work,
    and women's labour 191

households,
  and IS women 275
  and the private domain 219
  and women's work 222
  assets of 155
  budgeting for 257
  contradictions within 338–9
  survival strategies 181
  women's empowerment in 305–8
housework, 255
  and social reproduction 121–4
  intensity of 160–1
housing,
  female urban 233, 241
human capital,
  and childbearing 67
hunger,
  and women 32
husband's rule, 40, 78, 80
husbands,
  sharing women's work 103–4

ideology,
  and capital accumulation 205
  and female labour 191
  and hegemony 225
  and new production relations 220
  and sex difference in wages 193, 195
  and social reproduction 117–18, 124
  and women's condition 256
  and women's work 196–8
  gender 158, 167–70, 182
  Islamic 327–8
  of women's dependency 332
ignorance,
  women's sexual 33–44
illiteracy,
  female 41, 72, 162
  in Bangladesh 239

illness,
  children's 108
  female 31, 57
IMF policies,
  and Bangladesh economy 327
imperialism,
  and production relations 127
import substitution,
  retreat from 227
in-laws,
  husband's relations with 75–6, 82
income,
  diversification of 239–40
  women's 285
income generation,
  new recommendations for 315
  rural schemes for 346
  schemes for 284, 296–7
India,
  regional demography 132–8
Indian history,
  and gender 126–38
Indira Gandhi,
  downfall of 26
industrial estates,
  failure of 225–6
industrial growth,
  inadequate in Sri Lanka 262
industrial sector,
  in Bangladesh 216
industries,
  supported 227
inequality,
  ameliorating 169
  economic and social 155
  effects of development on 322
  gender 229
  in organizations 311–12
  in urban employment 240
  relations of 220
  reproduction of 52, 235, 241

infant mortality, 128
   and poverty 42, 165
   Indian historical 132–3
infanticide, 24–5
informal sector,
   women in 255, 263, 273–315
informal sector workers 13
infrastructure,
   and wage work for women 344
   development of 155
   technical 226
   urban 228
inheritance,
   and capital accumulation 124–5
intensification,
   and women 179, 203
   economic 12
   financial 196
intergenerational issues,
   and gender 17
investigation,
   of wage cheating 345
investment,
   in agricultural inputs 199
   policies to encourage 218
   public infrastructure 268
   rural industrial 226
irrigated agriculture,
   and women's work 189–90
irrigation,
   in Andhra Pradesh 183–5
IS organizations,
   membership in 280
IS women workers,
   recommendations to government 314
Islam,
   and changing norms 241
   and state policy 328–9
   and the Ershad regime 225
   in Bangladesh 216
isolation,
   breaking down of 348

in domestic sphere 25, 95
in the informal sector 274

Jats, 29
Jeffery, R. and P.M.,
   introduction 11
joint family,
   effects for new mothers 45
   relationships in 86–7
   work sharing in 104

Kerala,
   and basic needs 162
kinship,
   and women 5, 6, 16
   effects of working women on 230
   employment based on 232
kinship networks,
   and women in office 260
kinship systems,
   regional 71
knowledge of menstruation,
   lack of 33–5
Kolis,
   in fishing industry 277–82

labour,
   cheap 228
   reserve army of 216–17, 225
labour demand,
   agricultural 220
   and female disadvantage 195
   for women 190
   in capitalist agriculture 196
   regional differences in 334–5
labour disputes,
   union role in 288–9
labour force,
   and the state 217–18
   composition of in Bangladesh 235–6
   female, in India 154

INDEX

female, in Sri Lanka 254
reproduction of 51, 69, 116, 129, 200, 270
social construction of 216–18
stratification of 148
urban 215
women's role in 202
labour inputs,
  in millet and rice production 182
labour legislation,
  and IS workers 289
  in Sri Lanka 252
  recommendations for 314
labour market,
  gender segmented 12, 148, 229
  in urban EPZs 216
labour pool,
  role of kinship 232
  structure and norms of 217
labour power,
  reproduction of 121
labour process,
  fragmentation of 227
labour relationships, 110
labour surplus,
  effect on organizations 294, 310
land alienation,
  and women 128
land fragmentation,
  and fertility 67, 111
land reform,
  and reproduction 53
land sharing,
  by brothers 86
land tenure,
  and demography 137
  and women's status 127–8
landholding,
  and agricultural benefits 204
  and employment in EPZs 235–6
  and employment demand 221
  and fertility 56–7
  and rural social structure 336–7
  and women's work 196–8
  and women's condition 249
  by agricultural zone 182–3
  in canal villages 186
  in Shankpur 29
  male control over 202
landlessness,
  and gender relations 150–4
  and women 153–6
  and women's work 200
  effects of increases in 227
  in Bangladesh 220
leadership,
  development of women's 274, 293, 313
  male 282
  single-person 295
  Sri Lankan women in 258–9
  women and 334
legislation,
  deficiencies of 296–7
lending programmes,
  by banks 295
liberal temper,
  Sri Lanka's 247
life cycle,
  women's 110
life expectancy,
  in Sri Lanka 253
literacy,
  campaigns for 331
  female 253
loans,
  for women 284–7
  sex differentials in 332
  women and 334
low castes,
  condition of 154–5
low-skill jobs,
  and educated women 264–9
  and women 15

Madras Presidency,
  demography 132–8
mahila mandals,
  programmes of 293
male alliances,
  and women's work 197–9
male domination,
  and reform tradition 329
  and female dependency 25
  and male solidarity 338
Malthus,
  and Boserup 127
management,
  labour's loyalty to 233
manufacturing,
  and the global assembly line 227
  eldest children in 239
  male employment in 237
marital fertility,
  Indian 135, 142
marital problems, 75–8
market crops,
  grown by women 199
market economies,
  and women 3, 7
marketing,
  of handicrafts 226
  women's role in 222
marketing co-operatives,
  for women 315
marriage,
  age at 164, 247
  and changing norms 224
  and menarche 33–5
  aspirations for daughters 306–7
  Buddhist 248
  commercialization of 336
  for employed women 230
  into distant villages 79–80
  into nearby villages 79
  kinds of 74–82
  love 230
  studies of 5
  timing of 36
  upwardly mobile 150, 157
  with bought brides 81
  within same village 77
marriage age,
  and women's subordination 335
  female 72
  in Rajasthan 58
  in Sri Lanka 254
marriage networks,
  Hindu 78
marriage patterns,
  Muslim 77–9
married women,
  as workers 256
Marx,
  and laws of population 130
  and reproduction 24
  compared with Malthus 52–3
Marxist feminism,
  literature of 116–18
McCarthy, Florence 13
mechanization,
  effect on women 180, 190
medical care, 128
  of children 108
medical profession,
  and population policy 27
membership,
  in IS organizations 279
men's work,
  in agriculture 188–90
menarche, 33–5
menstruation,
  knowledge of 33–5
middle class,
  child rearing strategy 157
  in Bangladesh 222–4
  values regarding women 329
middle class women, 152–3
  and urban employment 236

education of 163–4
  in Sri Lanka 255–6
  occupations of 158
  political activity of 250–1
midwifery,
  village use of 43–4
migrants,
  female 263, 266
  use of 195
migration, 234
  costs of 162
  urban 228
militance,
  of rural Bangladesh women 342–7
  of women 258, 283
mobilization,
  and women's association 341
  electoral 257
  of rural women 323
  of women 13, 14
modern sector,
  and marriage arrangements 230
  women's employment in 225
modernization,
  agricultural 179
  and exploitation 148
  and Sri Lankan women 247
  and women 12, 158
  theory 14, 180
modes of production,
  and colonialism 130
  and gender inequality 124
  and population growth 127, 139
  pre-capitalist 119–20
modesty,
  and women's subordination 336
  female 35
monsoon,
  south-west 185

mortality,
  and gender 131
  and population growth 131
  and women's status 129
  in India 132
  sex differentials in 132–3
mother-in-law,
  role during childbearing 96, 98, 104
  role of 67, 73–4, 86–7, 91
motherhood,
  conditions of 100–9
mothers,
  and daughters 39, 53, 69, 84, 98–100
movement politics,
  and women 250
multilateral organizations, 327
multinationals,
  and women's work 160
Muslim women,
  conditions of 109
Muslim,
  and gender ideology 332
  as entrepreneurs 232–3
  in Bangladesh 222
  in Bijnor 70
  in Shankpur 29

natal kin,
  women's access to 72–3, 75, 79, 82
natal village,
  visits to 85, 91, 94, 97, 100
  women's behaviour in 72
nationalism,
  19th century 329
  and women 250
  and women's activism 330
net reproduction rates,
  Indian 132–4
networks,
  migrants' 235

of social control 337
of support among rural women 341
New Industrial Policy, 228
new mothers,
    care of 99–107
norms,
    changing 231, 254
    regarding women 266
nuclear family,
    effects for new mothers 45
    incidence of 158
nutrition,
    women's 28, 31–2, 44, 165–6
occupations,
    female 255
    gender characteristics 275
    generational mobility of 237
old age security, 74
    fertility and 167
oppression,
    and domestic sphere 25
    interdependency of 169
    patriarchal 147
organization,
    by women 152, 169–70, 250, 268
    of women by political parties 257
organizations,
    all-woman 332
    failures of 303–4
    occupational 273–4
    protecting female activists 343
outwork,
    kinds of 290–1
    manufacturing 278
overwork,
    effects of 31

pace of change,
    in Bangladesh 223
Pakistan,
    refugees from 29

parents,
    women's relations with 99, 108
parliament,
    women in 259
participation,
    and class 334
    drawbacks for women 313
    of women in labour force 254
    of women in organizations 297–308
    women's 322
Partition,
    migration effects 29
paternalism,
    in government policy 331
paternity,
    and property 124
patriarchy, 119–20, 146–9
    and agriculture 179–205
    and class exploitation 205
    and demography 158
    and development programmes 324
    and elite women's work 198
    and labour recruitment 232
    and women's employment 158
    in the workplace 215
    reproduction of 256
patrilocality,
    North Indian 68, 72
patronage,
    decline of 338
    transformation of 194–5
peasant households,
    and colonialism 130
    deterioration of 249
    fertility of 166–7
    status of women in 222
piecework,
    and IS women 278
    women's 160, 193, 290–1
plantation economy,
    export-led 249

INDEX

heyday of 252
ploughing,
  men's responsibility for 188–9
political action,
  and gender 13
political character,
  and culture 247
political domain,
  women's absence from 202
  men's control of 219
  upper class women in 248–9
political economy,
  and regional demography 137
  as a perspective 1
  South Asian 5
political mobilization, 7, 13
political office,
  women in 258–9
political parties,
  and IS organizations 295, 297, 310–11
  and women's interests 261
  in Sri Lanka 256
politicization,
  of women 167, 340
pollution,
  childbirth 99, 106
  during menstruation 35
  water 311
poor women,
  mobilization of 343, 344
  reproductive conditions 24
  self-help groups 342
population,
  and class formation 111
  and development 130
  and poverty 56
  laws of 130
  well-being of 252–3
population control,
  and women 26
  burden on women 339
  policy 327

population density,
  and colonialism 127
  and irrigation 183
  by agricultural zone 182
population growth, 14, 15, 26
  and capitalism 127
  and colonialism, 127–8, 130
  and gender 131
  and modes of production 127
  and poverty 130
  and proletarianization 56, 128
  components of 134–6
  in history 125
  in Sri Lanka 254
  lower class 52
  women as key to 331
population policy,
  and medical profession 27
  and women's status 27
  in Brazil 53
population pressure,
  in Bangladesh 223
  on land 111
poverty,
  and co-operatives 310
  and intensification 203
  and gender 156
  and illiteracy 163
  and population 56, 130
  and women's workload 160–1
  and women's status 128, 154–7
  demographic effects of 165–6
  increases in 325
  increasing in Bangladesh 223
  urban 228
power,
  and sexuality 39–40
  Mrs. Bandaranaike's 261
powerlessness,
  female 203
pregnancy,
  and health 28, 42, 45, 51, 128–9

of working women 264
women's work during 94–6
prices,
    relative to wages 205
private sector employment,
    for women 223
privatization,
    of women 115
production relations,
    creation of 217
    gender-based 229
    precapitalist 168, 249
    reconstitution of 241
    transformation of 190–1
    urban 216
    women's changing 223
productive forces,
    level of 248
proletarianization, 146, 240
    in Bangladesh 224
    of women 158
property,
    and gender inequality 124
    and generational mobility 125
    and women 71–2
public sphere,
    women in 247, 250, 256
purdah, 74, 239, 332, 338
    and urban labour markets 216
    Bangladeshi discourse on 230–1

quality,
    of employment 156–7
queens,
    in Sri Lankan history 249
quotas,
    and industrial development 227
rainfed agriculture,
    and women's work 188
Rajasthan, 11, 24–61
    village women 28–50
Ramamurthy, Priti 12

rape,
    by security forces 268
    investigation of 293
    wedding night 58
recruitment,
    through *maleks* 232
reform tradition,
    and male dominance 329
regional demography, 138
regional perspective, 9
relations of production,
    agricultural 180
    and colonialism 130
    and gender 140
    under patrons 194
reproduction,
    and class 129
    and freedom 24
    and gender 11, 26
    and land tenure 67
    and the family wage 224
    and widowhood 135
    and women's work 218
    as a conceptual problem 115
    biological 117, 120–2
    biological and social 127
    burden of 136–8
    in Bijnor, U.P. 11
    in Rajasthan 11
    individual 240
    of deprivation 161
    of industrial sector 217
    of inequality 52
    of labour power 50–1
    of patriarchy 120
    of social formations 52, 117
    of the labour force 200
    private sphere's burden 266
    state management of 26
reproduction rates,
    Indian 142
reproductive health,
    women's 15, 28, 97–109

resistance,
  forms of 335
  women's 322–3
resources,
  allocated to women's issues 328
  and gender 224
  and organizational success 280–1, 309–10
  and women's activism 334
  common property 155, 161
  degradation of 111
  government 325
  intergenerational transfer of 239
  scarcity of in Sri Lanka 253
  unequal distribution of 26
restructuring,
  global economic 225, 227
  in Bangladesh 216, 322
  of female workforce 26
rice cultivation,
  and women 182–3
rural economy,
  worsening conditions in 252
  transformation of 216
rural manufacturing,
  inadequate development of 224
rural population,
  and EPZs 240
rural women,
  and credit 226–7
  in Bangladesh 222
  mobilization of 323
  redefining 349–50
safety,
  and workforce control 234
  women organizing for 347
salaried work,
  sex gap in 157
Sangari and Vaid 8
sanitary protection,
  village women's 33–5
savings accounts,
  access to 339
school attendance,
  for girls 30
  in Kerala 162
seclusion,
  and female labour 184
  and urban labour markets 216
  of women in household 73, 124, 336
self-employment, 275–7
  among women in Bangladesh 227
semi-skilled jobs,
  women in 164
separation,
  from joint family 88–92, 97
sex of child,
  effect of 105
sex ratios, 28, 57, 186
  Indian regional 132–6
sexual access,
  to wives 40
sexual control,
  and gender 122
sexual harassment,
  by elite males 337
sexual intercourse,
  women's ignorance of 36–9
sexuality,
  and hierarchy 150
  and oppression 169
  and power 39–41
  and the division of labour 118
  control of 53
shame,
  during childbearing 94–100
sharam
  as modesty 35
  as shame 95–7
Sharma, Miriam 11
Shastri, Amita 13

shelter,
    women's 287
sisters-in-law,
    role of 102–5
skills,
    as assets 167
    required in EPZs 239
    women's employment 225
    women's improved 339
slowdown,
    of economic growth 267
slums,
    Bombay 277–8
social construction,
    of emerging labour force 215
    of women by reform movements 323
    progressive 169, 324, 350
social demography,
    and Indian history 11
social formations,
    reproduction of 51, 117, 139
    women's location within 335
social movements,
    and exploitation 333
    tradition of 329
social relations,
    of reproduction 51
social reproduction 67, 116–25
    and history 125–7, 139–40
    and housework 121–3
socialism,
    and women 125
socialization,
    and gender inequality 241–2
    for class roles 8
solidarity,
    among males 338
    and class divisions 12
    as a programme goal 332
    forging 168
    growth of among women 341
    kinship-based 234

    lack of among women 110, 124, 251
    women's 152–3
son-in-law,
    living in wife's home 74–7
son-preference, 109
    and women's status 336
    fertility effects 42
South Asian studies,
    and gender 4–5, 16
speedup,
    for women in agriculture 204
Sri Lanka, 13
    and welfare politics 251–3
    women in 246–70
state,
    and labour force creation 217–8
    and self-employed workers 281
    colonial organization of 249
    welfare politics of 252
    woman head of 246
    women and the 27, 74
status construction,
    and non-farm employment 222
status of women,
    and age at marriage 335
    and capitalism 127
    and Indian inferiority 330
    and reproduction 116
    in Sri Lanka 248, 261
sterilization,
    campaign for 26, 340
    of women 48–9
sterilization camps,
    conditions in 340
stratification,
    by zone 182
    labour force 148
    occupational 312
structural adjustment 227–8
structure,
    and cultural practice 218

INDEX

and women's opportunities 248
varying meanings of 8
of domination 325
subaltern studies, 333
and gender 6
subcontracters,
women 278, 290
subordination,
and capitalist patriarchy 184–5
and colonialism 329
and modernization 181
of women 109–11, 118–19, 124, 200, 248, 269
sexual basis of 41
to mother-in-law 93
subsistence,
and traditional relations 192
and wage levels 194
family strategies 337
women's contribution to 198
subsistence crisis,
and irrigation 183
subsistence holdings,
decline of 249
subsistence production,
and wage needs 224
and women's domestic work 218–19
liberation from 229
women's responsibility for 160–1
suffrage,
universal 251
women's 153, 246
survey research 274–9
survival,
and ethnic violence 268
and women's initiative 341
female 131, 253
sweepers,
union of 283, 289, 296
systems studies,

and women 10

Tamil militants,
in Sri Lanka 258, 267
technology,
agricultural 67
terms of trade,
deterioration in 252
textile industry,
decline in 311
tourist trade,
women in 262–3
trade reform,
in Bangladesh 220, 228
traditional relations,
breakdown of 338
transformation of 220
transformation,
of rural relations 325
tribals,
and landlessness 154
and wet rice cultivation 183
oppression of 151

unemployment,
and labour force creation 218
and political alienation 258
female 161
in Bangladesh agriculture 220
in Sri Lanka 263–5
rates by sex 256
unionization,
of female workers 169
unions,
and IS workers 273
curtailed 264
demonstrations by 289
malfeasance of 344
state limits on 228
women's participation in 298
United Provinces,
demography 132–8

unpaid labour,
  of women 121
unskilled labour,
  and women 25
untouchables,
  and agricultural labour 186
  and IS work 277
  in Shankpur 29
upward mobility,
  and kinship connections 233
  by sex 239
  constraints on in EPZs 229
urban areas,
  women alone in 230
  rural links of families 232
  female labour force in 224
  politicized 252
  occupational variation 280

vasectomy, 26
veiling,
  new uses of 231
violence,
  domestic 287, 291
  family 340
  female 342
  in Sri Lanka 262, 270
voting behaviour,
  of women in Sri Lanka 246

wage labour,
  agricultural 153–4, 221
  and fertility 165
  low skill 228
  women and 12, 149–50, 155, 191–2
wages,
  cash 193–4
  demand for 345
  family 224
  female 192–3, 233–4, 265
  in kind 194
  in rural economy 337
  increased need for 222
  increases in 203
  low 227
  minimum 228, 264
  of women in agriculture 335
  relative to prices 205
  sex differentials in 193–6
war,
  effects on women 268–70
water,
  and Green Revolution 29
  delivery contracts for 346
  from canals 185
  insufficient supply of 226
  pollution
  scarcity of 187–8
  collection by women 30, 161
welfare,
  politics of in Sri Lanka 251–3
West Bengal,
  employment in 157
wet rice areas 180–81
white collar work,
  women in 157, 164
widowhood,
  and reproduction 135
wife-beating, 39, 75, 80, 82, 86, 338, 347
  and the AMM 293
women,
  and agency 349–50
  and class 68, 72–4, 128
  and defiance 348
  and household cash 200–1
  and militant protest 342–7
  and open economy 262
  and participation 297–303
  and political activism 250
  and the police 343
  and the double day 204
  and the state 68, 72–4
  and the political domain 202
  and welfare gains 252–3

as workers 66–8, 71, 263
  in agriculture 186
  in higher education 253
  in industrial estates 226
  in politics 247
  in public life 267
  middle class 152–3
  political importance of 257–8
  politicization of 340
  protecting each other 347
  working class 273–80
women as workers,
  marginalization of 181
  projects for 294
women in development,
  studies of 2, 3
women's contributions,
  devaluation of 202, 329, 331
women's groups,
  development of 342
women's history,
  and the history of society 140
women's interests,
  and politics 261
  lack of representation of 327, 340
women's movement,
  in India 283
  radicalization of 340
women's roles,
  and female uplift 329
women's status,
  and population policy 27
  and poverty 129
women's studies, 7, 9
women's work,
  and cash requirements 196
  and elite social norms 198
  and male alliances 197–9
  and subordination 110
  as captive labour force 205
  biological 128–9, 133
  determinants of 155
  devaluation of 50–1, 70, 73, 121–2, 184, 219
  hours 229
  in agriculture 181–3, 188–90
  in construction 262
  in road building 344
  in the home 27–8, 83–4
  industrial outwork 290–1
  intensity of 30
  number of days 195
  reproductive vs. productive 218
  valorization of 122
work,
  public vs. private 219
work force,
  educated 238
  female 263–4
working class women,
  in Bombay 273–315
working conditions,
  in EPZs 229
  in FTZs 264
  of casual workers 281
  union efforts to improve 296
workload, women's 88, 90–2, 94–6, 105, 155–6, 160–2, 166, 179, 184, 195
  after delivery 99–100
  and financial intensification 203
  and irrigation 190
  and landlessness 156
world system,
  reproduction of 120
  and regional demography 137
  intensification in 12
  restructuring of 225
  social history 130
worries,
  women's 32, 269–70